Functional Safety and Proof of Compliance

Thor Myklebust • Tor Stålhane

Functional Safety and Proof of Compliance

 Springer

Thor Myklebust
SINTEF Digital
Trondheim, Norway

Tor Stålhane
Department of Computer Science
NTNU
Trondheim, Norway

ISBN 978-3-030-86151-3   ISBN 978-3-030-86152-0 (eBook)
https://doi.org/10.1007/978-3-030-86152-0

This Springer imprint is published by the registered company Springer Nature Switzerland AG.
The registered company address is: Gewerbestrasse 11, 6330 Cham, Switzerland

Safety does not happen by accident
Unknown

If you think compliance is expensive—
Try non-compliance
Former U.S. Deputy Attorney General Paul McNulty

Preface

On the top level, proof of compliance (PoC) is simple. A contract or a standard requires you to do something during development – in our case, the development of safety-critical systems and products. The PoC is a paper trail showing beyond any reasonable doubt that you have done what is required. In short, a PoC is used to build trust.

The material in this book is adapted to the IEC 61508:2010 series since this is a generic standard that several manufacturers use. In addition, we have included work related to ISO 26262:2018 – automotive – and EN 50128:2011/A2:2020 – railway. Other standards are used and referred to where they add useful information. Functional safety standards such as the ISO 13849 series and IEC 62061:2021 – machine safety standards – and DO-178/DO-254:2011 – aviation safety – are not derived from IEC 61508. Still, this book is also helpful for the manufacturers using these standards. In addition, the book is useful for operators using products and systems developed according to these safety standards.

PoC is important for the assessor and certification bodies when called up to confirm that the manufacturer has developed a software system according to the required safety and software standards. The PoC documents do not add functionality to the product. However, they add confidence, conditions, and trust to the product and simplifies certification, and as such, it is important for the product's value. Even so, the documentation needed for PoC is often developed late in the project and often in a haphazard manner.

We suggest an agile approach, and as a consequence of this, the majority of the PoC documents will be living documents, i.e., a document that is frequently edited and updated during the development process.

Using an agile development process creates several opportunities for improving the PoC process. The most important issue is that agile development with its daily standups will improve project communication. Improved communication will reduce the need for project-internal documents, and the manufacturer will produce only the documents that the assessor or the customer needs. The agile process will also improve communication between developers and customers. The PoC documents developed during the project will help to keep the customer up-to-date, show the customer that the project is on track, and allow the customer to correct the project's course when needed. In addition, only documents really wanted by the customer or assessor will be produced.

In agile development, developers do not take all the important decisions at the start of the project. Instead, they are taken when they are needed and the necessary information is available. The assessor should be involved when necessary. Thus, instead of discovering at the end of the project whether a PoC document is acceptable or not, this can be checked out when the specific job is done. It is always possible to ask the assessor, "If I do this, will this be accepted as PoC?" This will help developers produce the right documents. Note – you cannot keep the assessor hostage to your decisions, but he can tell you if he will accept it. He can also reject it for any reason. Due to the changing technology and changing operating environments, we foresee more changes to systems in the future. The influence of DevOps, which enables rapid feedback of errors and the need for new functionality from the system's operators to the developers, will increase this trend. This will increase the need both to update the safety cases and the PoC documents. In addition, we will need to perform change impact analyses for the changes. Thus, the developers need a standard for this and proof that it is done as required – more PoC.

This book's main audience is developers, assessors, certification bodies, and purchasers of safety instrumented systems.

Documenting the personnel's knowledge and experience is an important part of PoC. Just saying that the developers have performed an activity does not create all that much confidence. As a minimum, we need to have the following information:

- Education – university, courses
- Experience, including previous employers and assignments and hands-on experience with any tools and methods used in the process. Relevant dates should accompany the experiences.
- The resources – time and tools – are made available for the activities the developer performs during the project.

It is important to be aware that this should not be used by the management as an opportunity to put the blame for any problem on the individual developer. Whoever did the job, all responsibility still stays with the company's management.

PoC is an important input to the safety case since it is used to verify that the required activities have been done as stated and with sufficient quality. Just as the safety case, the PoC documents can also be built successively as the project develops. Building the PoC documentation and the safety case in parallel allows a stepwise acceptance of both PoC and the safety case. Thus, when the project is finished, the documentation is also finished. We avoid the long period after development where we wait for the assessor to finish the assessment work before the product can be shipped. When writing a safety case or a PoC, it is important to understand what the reader, e.g., the assessors, expect from the documents. This should be discussed and agreed upon with the assessor at the start of the project. Knowing what the assessor wants right from the start will enable the project to write safety cases and PoCs that are quickly accepted and thus do not slow down project progress. The safety case developers will profit from communication with those who create the PoC documents since they may need PoC documents for the safety case that may not be related to a customer requirement or standard requirement.

The book is organized as follows:

Chapter 1: "The Introduction" includes information related to proof of compliance in general, aspects related to communication between stakeholders, different standards, market access, and the link to the safety case.

Chapter 2, "Agile Practices," includes information related to benefits and reasons for doing agile development of safety-critical software, information related to agile practices, and related challenges and their agile solutions. This is an important alternative view on what to do instead of just looking at a set of practices. This chapter also includes requirements management and information related to alongside engineering (RAMS team). Finally, some practices are extended by the authors of this book to accommodate safety aspects.

Chapter 3, "PoC in Agile Development and for SMEs," includes information related to Agile and Kanban and development needed when having small SafeScrum teams. We discuss what is an argument and look into PoC for SMEs.

Chapter 4, "Generic Documents," includes information related to documents and information management, as well as practical advice related to living documents. We also discuss change impact analysis, code baseline, and configuration management. Last but not least, we discuss safety techniques and measures.

Chapter 5, "Plans and Functional Safety Management," includes information related to safety plans, functional safety management, and software quality assurance plans.

Chapter 6, "Safety Analysis Methods Applied to Software," includes information related to several safety analysis methods applied to software, common mode and common cause failures, and dynamic risks.

Chapter 7, "Safety and Risk Documents," includes information related to hazard logs (hazard record) and their relations to safety standards, safety, hazard, and risk analysis reports.

Chapter 8, "Software Documents," includes information related to the tool validation plans and tool processes. In addition, we have a look at release notes and change logs plus a subchapter regarding software architecture.

Chapter 9, "Test, Analysis and V&V," contains information related to test specifications, scripts, and reports plus software and hardware integration. The chapter also discusses the software quality assurance verification report and the architecture and design verification reports, software requirements verification reports, and the overall software test reports and software validation reports.

Annex, "Overview of Documents and Work Products Mentioned in Functional Safety Standards Including Weak Parts of Safety Standards," includes information related to documents or work products mentioned in relevant safety standards. We discuss documents not mentioned in safety standard series but often used by several manufacturers and weak parts of safety standards related to documentation and work products.

Trondheim, Norway   Thor Myklebust
                                                                                     Tor Stålhane

Acknowledgment

The authors thank the International Electrotechnical Commission (IEC) for permission to reproduce Information from its International Standards. All such extracts are copyright of IEC Geneva, Switzerland. All rights reserved. Further information on the IEC is available from www.iec.ch. IEC has no responsibility for the placement and context in which the extracts and contents are reproduced by the author, nor is IEC in any way responsible for the other content or accuracy therein."

The authors thanks ISO (International Organization for Standardization) for permission to reproduce figures and tables.

We are grateful for the layout and editorial comments from Springer and for their effective and professional work.

Contents

Acronyms

AAMI	Association for the Advancement of Medical Instrumentation
ABS	Anti-lock Braking System/Automatic Braking System
AC	Application Condition
ACEA	European Automobile Manufacturers Association
AI	Artificial Intelligence
AFD	Analyze First Development
AHL	Agile Hazard Log
ALARP	As Low As Reasonable Practical
API	Application Programming Interface
APIS	Authorization for Placing In Service
ASIL	Automotive SIL
ATC	Automatic Train Control
ATP	Automatic Train Protection
ATCS	Automatic Train Control System
ATSS	Automatic Train Signal System
AUTOSAR	AUTomotive Open System ARchitecture
BDD	Behavior-Driven Development
BRUF	Big Requirements Up Front
CAN	Controller Area Network
CAR	Corrective Action Request
CAT	Customer Acceptance Testing
CAU	Compact Antenna Unit
CB	Certification Body
CBI	Computer-Based Interlocking
CBSS	Computer-Based Signalling System
CBTC	Communication-Based Train Control
CC	Control Command
CCA	Common Cause Analysis
CCS	Control, Command, and Signalling
CE	Contracting Entity

CENELEC	European Committee for Electrotechnical Standardization
CHazOp	Computer HazOp
CIA	Change Impact Analysis
CIAR	Change Impact Analysis Report
CM	Configuration Management
CM	Common Mode
CMI	Controller Machine Interface
CMP	Configuration Management Plan
CoC	Confirmation of Change
CoC	Certificate of Conformity
COMAH	Control of Major Accident Hazards Regulations
ConOps	Concept of Operation
COTS	Commercial Off The Shelf
CR	Change Request
CRC	Cyclic Redundancy Check
CSI	Common Safety Indicators
CSM	Common Safety Methods
CSM RA	Common Safety Method for Risk Evaluation and Assessment
CST	Common Safety Targets
CTC	Centralized Traffic Control
CTS	Communication Transmission System
CVR	Conformity Verification Report
CVS	Conformity Verification Specification
DAT	Development Acceptance Tests
DataOps	Data and Operation
DeBo	Designated Body
DevOps	Development and Operation
DFMEA	Design FMEA
DIA	Development Interface Agreement
DMI	Driver Machine Interface
DNVGL-RP	DNVGL Recommended Practice
DoD	Department of Defence
DoS	Definition of System
DPP	Data PreParation
DR	Data Recorder
DRU	Diagnostic Recorder Unit
DSM	Daily Scrum Meeting
EA	European Accreditation
EASA	European Union Aviation Safety Agency
EBL	Emergency Break Limit
EIRENE	European Integrated Radio Enhanced Network
EMC	ElectroMagnetic Compatibility
EN	European Norm
EOA	End Of Authority

ERA	European Railway Agency
ERRAC	The European Rail Research Advisory Council
ERTMS	European Rail Traffic Management System
ETCS	European Train Control System
EU	European Union
EUC	Equipment Under Control
EUROCAE	European Organization for Civil Aviation Electronics
FAA	Federal Aviation Administration
FAI	First Article Inspection
FAR	Fatal Accident Rate
FAT	Factory Acceptance Test
FATC	Full Automatic Train Control
FFFIS	Form-Fit Functional Interface Specification
FIS	Functional Interface Specification
FIT	Failures In Time
FMEA	Failure Mode Effect Analysis
FMECA	Failure Mode Effect Criticality Analysis
FMEDA	Failure Mode Effect Diagnosis Analysis
FRACAS	Fault Recording Analysis and Corrective Action System
FRS	Functional Requirement Specification
FSA	Functional Safety Assessment
FSM	Functional Safety Management
FS	Full Supervision
FTA	Fault Tree Analysis
GA	Generic Application
GALE	Globally At Least Equivalent
GAMAB	Globalement Au Moins Aussi Bon
GAME	Globalement Au Moins Equivalent
GASC	Generic Application Safety Case
GMA	General Morphological Analysis
GP	Generic Product
GPS	Global Positioning System
GPSC	Generic Product Safety Case
GRAC	Generic Risk Acceptability Criteria
GRS	General Requirement Specification
GSM	Global System for Mobile Communication
GUI	Graphic User Interface
HARA	Hazard Analysis and Risk Analysis
HazId	Hazard Identification
HazOp	Hazard and Operability Study
HL	Hazard Log
HR	Hazard Record
HS	Hot Standby
HW	Hardware

IA	Impact Analysis
IAC	Intermediate Application Condition
IAEA	International Atomic Energy Agency
IAF	International Accreditation Forum
IC	Interoperability Constituent
ICE	Integrated Clinical Environment
IEC	International Electrotechnical Commission
IECEE	IEC System of Conformity Assessment Schemes for Electrotechnical Equipment and Components
IEEE	Institute of Electrical and Electronics Engineers
IF-FMEA	InterFace FMEA
IID	Iterative and Incremental Development
IM	Interface Management
IMO	International Maritime Organization
INEA	Innovation and Networks Executive Agency
IRIS	International Railway Industry Standard
ISA	Independent Safety Assessor
ISAD	Infrastructure Support for Automated Driving
ICC	Interim Certificate of Conformity
ISV	Intermediate Statement of Verification
IT	Information Technology
ITC	Installation, Test, and Commissioning
ITSM	Information Technology Service Management
JRU	Juridical Recorder Unit
KMC	Key Management Center
KMS	Key Management System
LEU	Lineside Electronic Unit
LOOP	List Of Open Points
LOPA	Layers Of Protection Analysis
MA	Movement Authority
MAR	Movement Authority Request
MEM	Minimum Endogenous Mortality
MIL-Std	Military Standard
MIL-HDBK	Military Handbook
ML	Machine Learning
MLOPS	Machine Learning Operation
MVP	Minimum Viable Product
NANDO	New Approach Notified and Designated Organisations Information System
NASA	National Aeronautics and Space Administration
NCR	Non-Conformity Report
NCR	Non-Compliance Report
NNSR	National Notified Safety Rules
NNTR	National Notified Technical Rules

NoBo	Notified Body
NOROG	NORsk Olje og Gass
NSA	National Safety Authority
NSR	National Safety Rules
NTR	National Technical Rules
OBU	On Board Unit
OC	Object Controller
ODD	Operational Design Domain
OEDR	Object and Event Detection and Recognition
OEM	Original Equipment Manufacturer
OJEU	*Official Journal of the European Union*
ORA	Operational Risk Analysis
OS	On Sight (operational mode)
OTA	Over The Air
O&SHA	Occupational and Safety and Health Administration
PaaS	Platform as a Service
PDCA	Plan Do Check Act
PHA	Preliminary Hazard Analysis
PIU	Proven In Use
PLL	Possible Loss of Life
PoC	Proof of Compliance
PR	Passive Recommendation
PTC	Positive Train Control
QA	Quality Assurance
QMR	Quality Management Report
R	Recommendation
RA	Railway Authority
RAC	Risk Acceptance Criterion/Code
RAM	Reliability, Availability, and Maintainability
RAMS	Reliability, Availability, Maintainability, and Safety
RAMSS	Reliability, Availability, Maintainability, Safety, and Security
RBC	Radio Block Center
RFU	Recommendation For Use
RHA	Requirement Hazard Analysis
RPN	Risk Priority Number
RTCA	Radio Technical Commission for Aeronautics
SAD	Safety Architecture Description
SADT	Structured Analysis and Design Techniques
SAR	Safety Assessment Report
SART	Structured Analysis for Real Time
SASC	Specific Application Safety Case
SAT	Site Acceptance Test
SC	Safety Case
SCS	Safety-related Control Systems

SCSW	Safety-Critical SoftWare
SEI	Software Engineering Institute at Carnegie Mellon University
SEooC	Safety Element out of Context
SecRAC	Security RAC
SHA	Safety Hazard Analysis
SIL	Safety Integrity Level
SME	Small and Medium Enterprises
SMR	Safety Management Report
SMS	Safety Management System
SotIF	Safety of the Intended Functionality
SQAP	Software Quality Assurance Plans
SQT	Safety Qualification Tests
SRAC	Safety-Related Application Condition
SRS	System Requirements Specification
SRS	Safety Requirements Specification
STAMP	Systems-Theoretic Accident Model and Processes
STPA	Systems Theoretic Process Analysis
SwAR	Software Assessment Report
SW	SoftWare
SW-FMEA	SoftWare FMEA
T&M	Techniques & Measures
TASC	The Agile Safety Case
TCC	Traffic Control Center
TF	Technical File
TFD	Test First Development
TEN	Trans-European transport Network
THR	Tolerable Hazard Rate
TOGAF	The Open Group Architecture Framework
TSI	Technical Specification for Interoperability
TSP	Team Software Process
TSR	Technical Safety Report
TRF	Test Report Format
UL	Underwriters Laboratories
UML	Unified Modelling Language
UX	User Experience
V&V	Verification & Validation
WIP	Work In Progress
XP	eXtreme Programming
YAGNI	You Ain't Gonna Need It

Chapter 1
The Introduction

The Moving Finger writes; and, having writ,
Moves on: nor all your Piety nor Wit
Shall lure it back to cancel half a Line,Nor all your Tears
wash out a Word of it.

Omar Khayyám: The Rubáiyát

What This Chapter Is About
- Why proof of compliance
- The importance of communication
- Prescriptive and goal-based standards
- Market access
- The PoC link to the safety case

1.1 What Is This Book About?

As indicated by the title, this book is about proof of compliance to one or more safety standards. A lot of material is adapted to IEC 61508:2010 series since this is a generic standard that several manufacturers heavily use. In addition, we have included work related to ISO 26262:2018—automotive—and EN 50128:2011/A2:2020—railway. Other standards are used and referred to where they add useful information. Functional safety standards such as ISO 13849 series and IEC 62061:2021 (machine safety standards) and DO-178/DO-254:2011 (aviation safety) are not derived from IEC 61508. Still, this book is helpful for the manufacturers using those standards too. And the book is useful for operators using products and systems developed according to these safety standards.

Proof of compliance (PoC) is important for the assessor and certification bodies when called up to confirm that a software system has been developed according to the required safety standards. Even though the PoC documents do not add functionality to the product, they add confidence and trust to the product and ease certification, and as such, it is important for the product's value. In addition, PoC with

T. Myklebust, T. Stålhane, *Functional Safety and Proof of Compliance*,
https://doi.org/10.1007/978-3-030-86152-0_1

required safety standards is important in questions related to insurance and criminal liability cases. Unfortunately, the documentation needed for PoC is often developed late in the project and in a haphazard manner.

We suggest an agile approach. Consequently, the majority of the PoC documents will be living documents, i.e., continually updated during the development process. Therefore, we will focus on the following issues:

- Each of the defined activities needs to leave a "paper trail"—documentation that:

  - It has been done as required by the standard, or an alternative approach has been agreed upon with the safety assessor.
  - It has been done by engineers with the right competence and experience.
  - Sufficient resources—time, tools, etc.—have been made available for the job.

- Activities only needed to provide PoC must be included in, e.g., the project plan or documentation plan so that the activities are done by the manufacturer as soon as the necessary information is available.
- Improve communication between the stakeholders. Communication is especially important when we have a lot of living documents.
- All information needed for PoC must be written only once and referred to when needed somewhere else. This is important since we may need to change some of the PoC-relevant information later.
- Define templates for important documents such as test reports and code reviews. This will increase the opportunities for reuse of all or significant parts of relevant documents.

The mix of agile, PoC, and safety processes as demanded by standards is challenging. Our approach focuses on the main PoC documents and work products mentioned in safety standards like, e.g., EN 50128, ISO 26262, and Annex A in IEC 61508-1. They are not only mentioned in the safety standards but are also referenced in the safety cases. If a safety case has to be issued or the company wishes to issue a safety case for other reasons, these PoC documents must be mentioned in the safety case.

In this book, we have tried to follow the development process when writing about the PoC needs. The first three sections are an exception—first about agile development, a topic close to the authors' hearts; then on developing and using generic documents, a nice way to save work and improve efficiency; and finally about plans. To quote Mike Tyson: "Everybody has a plan until they get punched in the mouth."

Then we handle risk and safety analysis. All safety-critical projects start here. The operational risk will mostly decide the development process through assessment of SIL or ASIL or similar indicators in other application areas, e.g., performance level (PL) when developing machines (ISO 13849). When we have defined the development process, we will start with the architecture and software documentation. To finish the project in an orderly manner, we discuss proof of compliance related to testing, analysis, and Verification & Validation (V&V) in the last section.

Safety cases are important and will be even more important in the future. Several pieces of information that are important for a safety case can be found in this book—e.g.:

- The different plans, including the safety plan, see Sect. 5.1
- Change impact analysis in Sect. 4.3
- Quality assurance and release and deployment in Sect. 8.2
- The hazard log in Sect. 4.1
- The architecture in Chap. 6
- Test, analysis and verification, and validation in Chap. 8

The book ends with a list of acronyms used, just to clear up the alphabet soup used in the other sections.

1.2   Front Page of Documents

A front page of a PoC document should normally include:

- Logo of the responsible company
- Title
- Name of system, item, or component
- Version number of the system, item, or component
- Date of issue and revision number
- Document number
- Author(s)
- Reviewer(s)
- Approver(s)

1.3   Reuse of Documents

When you read through this book, you might be overwhelmed by the large amount of needed documentation. A lot of the needed documents are intended for the safety assessor, certifying bodies, or safety authorities. The first thing we need to keep in mind is that the assessors are not primarily looking for documents but for information. The second important thing is that we are here not talking about documents but about documented information. ISO 9001:2015 states:

> Where ISO 9001:2008 used the term 'records' to denote documents needed to provide evidence of conformity with requirements, this is now expressed as a requirement to 'retain documented information'. The organization is responsible for determining what documented information needs to be retained, the period of time for which it is retained and the media to be used for its retention.

Also note that ISO 9001 auditors are willing to accept whiteboard snapshots of the discussions if they include the date and a list of participants as proof of conformance for an activity. At least some IEC 61508 safety assessors have confirmed that they will do the same. In addition, the level of trust may affect the level of documentation required by the assessor. See also Sect. 9.4—Software quality assurance verification report.

New documents have a high cost since they have to be written more or less from scratch for each new project. It is therefore beneficial to make use of already available templates that have been published as industry papers, e.g., "change impact analysis" (Myklebust et al. 2014) and Sect. 3.3, or published by organizations developing guidelines such as Misra (2021) and AAMI (2020).

The following is a set of suggestions intended to reduce the amount of work related to PoC for a project.

- Reuse of the development process will create more document reuse opportunities.
- All information should only be registered once. Other documents that need this information should refer to the first occurrence. In this way, we are sure that any necessary updates will be available to all documents where it is needed.
- The company should have standard templates for all documents and reports, with clear guidelines for inserting new information. In addition, they need a procedure for how to change the template if this turns out to be necessary.
- The company needs a database with reusable documentation. This includes, but is not limited to, documentation of tools validation, report templates, methods, and personnel—trainees and experienced. In addition, the company needs procedures for updating the database contents.
- Structure all documents so that they are able to include or refer to relevant computer-generated information such as test results.
- All new documents should be constructed with reuse in mind. Thus, whenever we plan to include some information, we should ask "Is this information already available somewhere else?" This will help to build a reuse culture into the company.

An extensive overview of reusable documents plus suggested content can, e.g., be found in Anderson (2007).

Two things will help to increase reuse—build an on-line library of documents already developed and build a library of instruction on how any document needed in your process should be developed—including the necessary templates. This will increase the reuse opportunities.

According to Levy (1993), the four processes used to create new documents and new versions of old documents can be described using the four Cs (parts are rephrased to take into account modern data possibilities):

- Creation: the production of new material, accomplished for instance by inputting text using computers or writing on a piece of paper.
- Collection: the identification and gathering together of previously existing material—data, documents, or pieces of documents and, e.g., films.
- Combination: the stitching together of new and old material to form a new unity.
- Customization: the reworking of this material to fit its new setting. Collection, combination, and customization all involve reuse; only creation introduces new material.

Table 1.1 Approaches to document reuse

Approach	What is it	Form of reuse
Interchange	Copying material from one environment to another, e.g., from a personal computer to a shared system	Moves material, enabling collections and sharing. Dynamic document creation is also possible
Composite structure	Specifying how pieces can be put together to make new units	Enables combinations as well as extractions and re-combinations
Presentation independence	Representing material independent of presentational characteristics	The basis for transformation to customize to different physical realizations, e.g., from paper to scanned papers that the involved parties share to a common data system
Composite structure	Representing abstractions common to the various doc. representation schemes	Meta-customizations. Document metadata in Microsoft Word, includes, e.g., the file size, date of document creation, the names of the author(s) and most recent modifier, the dates of any changes, and the total edit time. In addition, editing features like the "track changes" (that is also important for the assessor) option in Word also generate metadata such as deleted text and, e.g., comments between authors

Table 1.1—adapted from Levy (1993)—also shows four approaches toward reuse of documents. A library reuse approach has been suggested by Taipuva (2018). The following quote gives a good idea of their concept, which is especially important when we claim compliance with a standard:

> When you derive requirements based on standards into a database of requirements, make sure the whole standard document is covered. You may, for example, create one requirement based on each numbered section of the standard. Requirements should also contain a reference to the original source, stating clearly the standard name and version number. A good practice is to create one requirement specification document and a corresponding test case specification document against one standard document.

We see from Fig. 1.1 that the decision to reuse one document can enable the reuse of several other related documents—e.g., reuse of the validation plan will open up for also reusing the validation report. Thus, structures like the one shown in Fig. 1.1 will be useful for a lot of reports and other documents.

As part of a study of relevant proof of compliance documentation when certifying products according to IEC 61508:2010, we found that more than 50% of the documents can be reused (Myklebust et al. 2014). These documents have to be made as generic as possible by the manufacturers. For documents that have to be updated over several sprints, reuse solutions is important. These documents could, e.g., include tables or point lists that are easily updated. Reusability of tests and analysis should also be included in these evaluations—see IEEE 1517:2010 for reuse processes for software. Reuse is also important if we want to perform regression tests automatically and effectively.

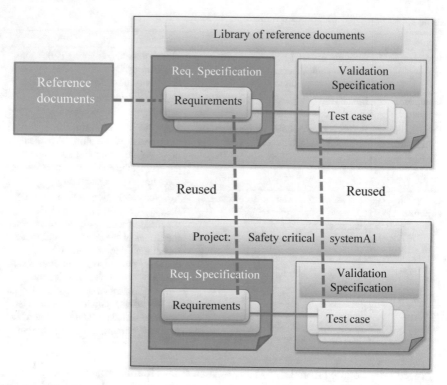

Fig. 1.1 Libraries and document reuse

Some standards, such as ISO/IEC/IEEE 29119-3:2013, include procedures and templates for reports such as Test status report, Test completion report, Test data readiness report, Test environment readiness report, and Test incident report. As part of the SafeScrum mindset, it is important to reduce the amount of documentation, and the assessor should be involved early in the project to discuss the relevant level of information to be delivered by the manufacturer to the assessor. What could be the minimum of documentation delivered to the assessor should therefore be discussed before starting to develop any new document. Some of the information could be reviewed by the assessor as part of audits and technical meetings.

1.4   Communication

1.4.1   Why Is Communication So Difficult

In non-safety-critical projects, we have two main types of communication—between the developers and the customer and between the project participants. For safety-

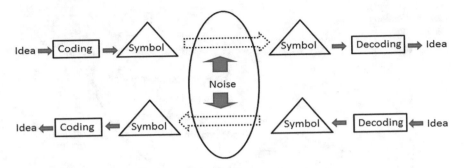

Fig. 1.2 Coding and decoding in person-to-person communication

critical projects, however, depending on the domain, we have to add a third category—between the developers and (1) the approval authority and (2) the certifying body. As we expose the public to more and more safety-critical applications—e.g., autonomous cars and busses—we will also see a fourth category—between application owners and the general public (see also Myklebust and Stålhane (2021)). In safety-critical projects, we need communication to clarify safety requirements, fix temporary problems, resolve conflicts and obstacles, and share safety knowledge (Wang et al. 2018).

All experience and research indicate that communication is important in all phases of a software project. Research suggests that software practitioners with good communication skills are critical for the success of software projects. Communication underpins other success factors such as customer involvement and good management. Together with interpersonal skills, good communication skills were ranked the third most important trait for exceptionally performing software developers in a survey of IT professionals (Wynekoop and Walz 1999). Communication is important at all stages of the project life cycle, both within a company and with external bodies.

Research shows that there is a positive link between internal task-related communication and performance of software development projects. On the other hand, lack of communication causes project failures since it can lead to lack of agreement on project objectives and targets prior to the starting developing a safety-critical system. A study of 21 software development teams showed that internal task-related communication accurately predicts overall software project performance. In addition it is a particularly accurate indicator when used during the late stages of project development (Brodbeck 2001).

There are at least six messages in a communication between two persons—from your side, what you mean to say, and what you really say. From the other side, what the other person thinks he hears, what the other person means to say in response, and what the other person really says in response. Finally, it is important that you believe what you hear. No wonder some people prefer communicating with computers. The authors have adapted the model shown in Fig. 1.2 from a model presented by eCampusOntario (2018).

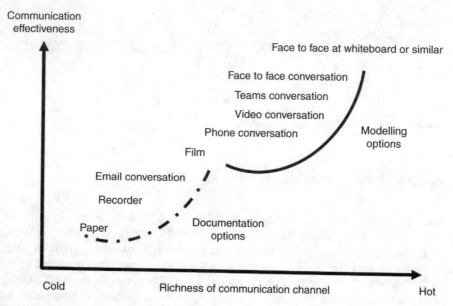

Fig. 1.3 Communication channel efficiency

In addition to the simple diagram of coding and decoding between two persons communicating, the channel through which they communicate is also important. The following diagram describes the efficiency of communication, depending on the communication channel used. The diagram is adapted and modernized from Ambler (2002) (Fig. 1.3).

More detailed information is summarized in Table 1.2. The Ambysoft survey (Ambler 2008) explored the concepts captured in Table 1.1. Ambler's comments are worth quoting: "The results are summarized below with answers rated on a range of −5 (very ineffective) to +5 (very effective). Note that overview documentation was perceived as being reasonably effective although detailed documentation was not. Also, online chat was thought to be effective between developers but not with stakeholders, likely a reflection of cultural differences and experiences between the two communities." The communication effectiveness is split between "within team" and "with stakeholders." As we see, the differences between the effectiveness in these two cases can be considerable—see Ambler (2002).

In addition to the communication mode, a second factor will also influence the effectiveness of the communication, namely, proximity—emotionally, temporally, and physically.

- Physical proximity. The closer people are to one another, the greater the opportunities for communication. At one end of the spectrum, two people can be working side-by-side and at the other end of the spectrum two people can be in different buildings.

Table 1.2 Efficiency of communication within teams and between team and stakeholders

Communication strategy	Within team	With stakeholders	Comments related to assessors
Face to face	4.25	4.06	Relevant at kick off and regular meetings
Face to face at a whiteboard	4.24	3.46	
Overview diagrams	2.54	1.89	Relevant depending on the system and the projects
Online chat	2.10	0.15	Important due to the use of Teams and similar tools
Overview documentation	1.84	1.86	Important part from day one of the project
Teleconference calls	1.42	1.51	Practice before the 2020 Corona pandemic and replaced by Teams or similar after the pandemic
Video conferencing	1.34	1.62	Practice after the 2020–2021 Corona pandemic
E-mail	1.08	1.32	Less used after the 2020–2021 Corona pandemic
Detailed documentation	−0.34	0.16	Main part of an assessor's work. In the future, we expect that more is performed through access to the documentation system

- Temporal proximity. Whether or not two people are working at the same time affects communication.
- Amicability—the willingness to listen with good intentions. Cockburn (2008) believes that amicability is an important success factor. The greater the amicability, the greater amount and quality of information will be communicated and less will be concealed.

Nothing beats face-to-face communication—preferably supported by a whiteboard or a large flip-over. Nothing is worse that communication through large amounts of detailed documents. However, for many companies, the preferred way of communicating with customers and especially with safety assessors is through detailed documents—the least effective way of communicating (Owens 2010). Face-to-face communication is best and paper is way down on the list. Over-the-wall engineering, which is far from efficient, is the most used way of communication during the development of large, complex, and safety-critical software systems. The main reason for this is the standards' focus on documents as the main results of any activity (Fig. 1.4).

The opposite of "over-the-wall" engineering is the tiger team. In the tiger team approach, a problem belongs to everybody in the project. Everybody communicates face to face and the focus is on the problem at hand. A tiger team is the closest traditional software development gets to being agile. The main idea of a tiger team is to involve everybody that can contribute to the problem's solution, e.g., customers, developers, designers, system architects, and testers—see also Lucidchart (2020) (Fig. 1.5).

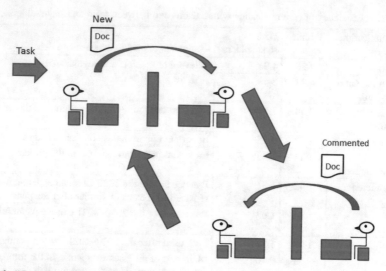

Fig. 1.4 "Over-the-wall" engineering

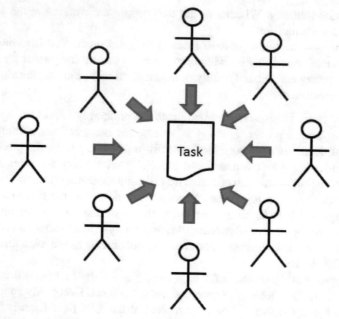

Fig. 1.5 The tiger team approach to problem solving

1.4.2 Communication in Agile Projects

The main advantage with agile development is communication and a continuous evaluation of the work done since communication gives confidence and comfort. For agile projects, the majority of communication in a safety-critical project takes place in a group setting.

- Within the project: daily stand-up meetings and retrospectives after each sprint. Meetings between the RAMS team and the Sprint team.
- With stakeholders: during safety analysis. The safety assessor does not participate here but will scrutinize the analysis report later.
- Between developers, management, and assessors during certification and certification-related activities.

For agile development to succeed, we need an efficient group—i.e., a group that is characterized by having a goal accepted by all participants and an open communication climate. In addition, the group members need to have confidence in each other and share interest, understanding, and recognition. All decisions should be made based on the whole group's opinions and last, but certainly not least, they need flexible management.

On the other hand, an inefficient group is characterized by tension, hostility, and indifference plus inefficient communication—all participants work only to get their opinion accepted. All critique is negative or embarrassing and creates tension, and decisions are mostly made through voting, where a simple majority is enough.

In our opinion, the main success factors when moving from plan-driven to agile are that it increases the quality of communication, the opportunities for communication, and the amount of communication. In addition, it allows us to move from over-the-wall engineering to tiger teams. In a survey of 120 companies, communication skills were among the three most important skills considered when hiring new developers—together with adaptability and software development skills (Stålhane et al. 2019).

Communication with the assessor is always important when developing safety-critical systems, and the assessor should thus be involved from day 1. It will make the whole process less costly if the project generates the necessary documents in a timely manner and lets the assessor comment on them as soon as they are ready. Representatives from the certifying body must be present at all meetings related to system safety. This will reduce the probability of making decisions that later will not be accepted by the certifying body. In addition, it will give the certifying body confidence in our work and give them the necessary context information. In compliance with the Administration Act's §11 a NoBo (Notifying Body) has a general duty to provide guidance. Through its guidance, the NoBo shall point out possible faults or shortcomings of a product so that the manufacturer can bring the product into line with the requirements from regulations. It is, however, the manufacturer's responsibility to find the technical solutions.

Based on interviews with assessors, we see that 80% of all problems during certification could have been solved early in the development by direct contact between the assessor and the project participants. The reasons for not doing so are to a large degree cultural. The main remaining challenge is "proof of compliance" to relevant development standards. The best solutions to the PoC problem are related to communication. In order to make it work, we need frequent meetings with the assessor. This will give both parties a better understanding of what is expected from the other and ample opportunities to clear up misunderstandings. An extra benefit is that familiarity and confidence with the company and the project gives the assessor more flexibility on what to accept as PoC.

1.5 Communication and Learning from Experience

For software development, as for most other trades, it is important to learn from experience which in agile development is done through person-to-person contacts and groups, such as retrospectives. An agile retrospective is a simplified version of Post-Mortem Analysis (PMA). In some sense, communication is the very essence of retrospectives and also of PMA. The idea is simple: use brainstorming to identify problems and improvement opportunities. Identify ways to fix problems and grab the opportunities to realize improvements. However, brainstorming is dependent on the participants' ideas, knowledge, and experience, which can deceive people. The following issues are important when we assess the results of a brainstorming session—see Kroeck et al. (1989):

- Selective perception—People develop a schema and a knowledge base through which they understand the world. This creates a filter through which all info must pass. Information is adapted to fit or verify what is expected.
- Anchoring—Earlier evaluation is an anchor. Insufficient information in the next evaluation result in the use of the mean value of previous evaluations. This can especially be dangerous when developing safety-critical systems.
- Concrete info—A single, vivid event tends to be much more important than more abstract events or statistical info.
- Personal experiences, even second hand, often have a greater effect than numerical data.
- Law of small numbers—People extrapolate from small samples to large populations without checking for significance. This can especially be dangerous when developing safety-critical systems.
- When people face a large volume of information, they tend to eliminate areas that are unclear or would require more information for a clear understanding.
- People focus their attention on what they consider to be the most likely hypothesis even though this outcome has not been tested or verified. This can especially be dangerous when developing safety-critical systems.

- They often assume that events and characteristics are correlated merely due to temporal occurrences or same category membership.

A large part of the problems related to memory is the long time from experience to info extraction. For agile projects, this can be anything from a week to a month. Material from stand-ups will help—e.g., snapshots of the whiteboards used for discussions. Thus, the majority of the documents used here should be snapshots from discussions on a whiteboard.

Any software development is critically dependent on communication. An important question to ask for any changed or new activity is thus: Will this improve communication? Everything that improves or increases communication is good while everything that hinders or reduces communication is bad. A person's ability to convey info in a face-to-face meeting is large. The ability to use memory and expertise to solve new problems is fantastic. However, there are some weaknesses that we need to remember and to cater for. Otherwise, we may be seriously misled.

As the Chaos report shows and all our experience confirms, communication is important. All projects that go down the drain do so because of two main reasons—bad management and bad communication. To underpin this, we will end with a quotation from a focus group in a Norwegian company. When discussing why developers make errors, one developer really got the attention when he said: "Everybody in a development team have the same information when the project starts. Each person's understanding of the system undergoes changes during the project. Unfortunately, the changes in understanding are not the same for each person."

As said above, your colleagues are an important source of information and experience. The work of Kroeck (1989) has identified some sources of uncertainty that need to be considered. Recently, Tim Harford (Harford 2021) has published an important book called *How to Make the World Add Up* with some observations that are a little disturbing:

> We often find ways to dismiss evidence that we don't like. And the opposite is true, too: when evidence seems to support our preconceptions, we are less likely to look too closely for flaws. It is not easy to master our emotions while assessing information that matters to us, not least because our emotions can lead us astray in different directions.

And then:

> Psychologists call this "motivated reasoning". Motivated reasoning is thinking through a topic with the aim, conscious or unconscious, of reaching a particular kind of conclusion. . . People with deeper expertise are better equipped to spot deception, but if they fall into the trap of motivated reasoning, they are able to muster more reasons to believe whatever they really wish to believe.

1.6 Communication and the Guilds

A guild is a community of practice, i.e., a group of people with a common set of ideas, knowledge, and interests but working in different project teams. In the guild slang, a team is called a squad—see Fig. 1.6. For large-scale agile development, the guild culture also introduces the two terms chapters and tribes (Šmite et al. 2019).

- Tribe: Contains 30–200 people who share a clear mission and a set of principles. A tribe has an experienced senior leader and has all the skills needed to develop software features end to end.
- Chapter: A group of engineers who have the same manager and are focused on personal growth and skills development. Engineers in each chapter share knowledge, learn from each other, and discuss common challenges.

In a SafeScrum context, the sprint team is a squad whereas engineering teams—e.g., the RAMS team—are guilds (see also Sect. 2.3 "Agile practices"). According to NewsCred (2021), there are a few things that need to be done in order to make the software guild idea work:

- The team needs to set aside time to do all of their work. NewsCred allocates a full day to every sprint for the team to focus on their guild work. When necessary, the RAMS teams should be included.
- Create a mechanism of accountability—the guild will be evaluated on the performance related to what they are responsible for. NewsCred suggests two ways to create accountability for the guilds:

  - Jira board—a Scrum board or a Kanban board.
  - Guild fest—all the guilds present their works to the entire tribe.

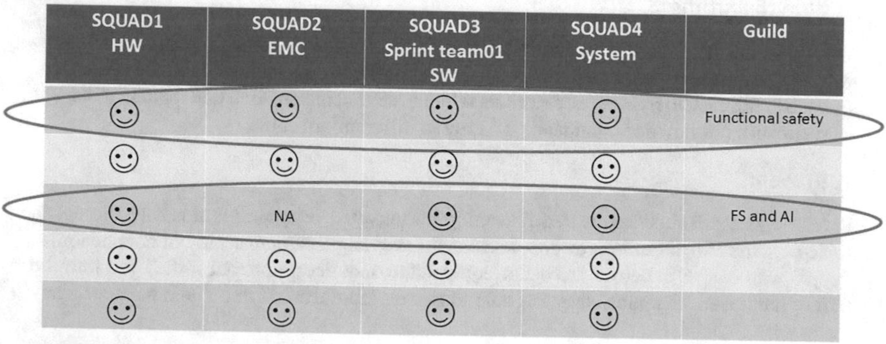

SQUAD1 HW	SQUAD2 EMC	SQUAD3 Sprint team01 SW	SQUAD4 System	Guild
☺	☺	☺	☺	Functional safety
☺	☺	☺	☺	
☺	NA	☺	☺	FS and AI
☺	☺	☺	☺	
☺	☺	☺	☺	

Fig. 1.6 Squads and guilds

- Be open to feedback and be prepared to change the process depending on the culture and operations of your team. This can affect both the Sprint team(s) and the RAMS team.
- Make sure that there is someone who manages the guilds. This person must be able to

    - Sell the idea of guilds to the team and the organization at large.
    - Own the process and guidelines for guilds.
    - Hold the guilds accountable and make sure they aren't underperforming.

The organization based on squads and guilds looks much like a matrix organization but the uses are different. In matrix organizations, people with similar skills are "pooled" together into functional departments and "assigned" to projects, and they "report to" a functional manager. The squads represent the vertical dimension while the horizontal dimension is for sharing knowledge, tools, and code (Kniberg and Ivarsson 2012).

The following recommendations for successful guilds are adapted from Šmite et al. (2019):

- Establishing a clear practice. Guilds that had disagreements about what a practice represents or did not have a clear direction tend to fail.
- Demonstrating mutual engagement. Problems related to engagement, attendance, and representative membership was due to a lack of dedicated time, low motivation, high turnover, and lack of colocation. The guild coordinator should try to overcome this by proactively contacting the members.
- Interacting regularly. Communications channels such as Slack and Google groups will provide interaction and transparency. However, face-to-face sessions are crucial for boosting guild activity.
- Improving practice. The majority of members recognize the guilds' ability to create value for both members. The top recognized benefits across the guilds are improved business outcomes and coordination across units.

In addition, the use of guilds improves the psycho-social environment by providing a sense of belonging and fun of being with colleagues and the ability to network and expand skills and expertise (Šmite et al. 2019).

Just as an afterthought, beware of group thinking—a psychological phenomenon that occurs within a group of people in which the desire for harmony or conformity in the group results in an irrational or dysfunctional decision-making outcome (Wikipedia 2021). In addition, Adam Smith emphasized the limitations imposed by the guilds and their detrimental effects on the economy. Karl Marx agreed that the guilds were one of the major impediments whose elimination would permit the development and implementation of the technological changes associated with the Industrial Revolution. When Adam Smith and Karl Marx reach the same conclusion based on the same evidence and reasoning, we should pay attention, because it doesn't happen all that often (Horn 2005).

1.7   Market Access and Certification

To capitalize on global opportunities, manufacturers developing safety-critical systems should seek to bring their product or system to the global market. Developing products and systems that meet all the leading markets' requirements is a rewarding and sustainable approach for manufacturers. A world of opportunities opens up when a product meets all the compliance requirements of the relevant markets for the product or system. To ensure quick market access, the following tips are important:

- Know the relevant and applicable standards.
- Discuss with experts, e.g., assessors that have the knowledge regarding the relevant countries and regions.
- Ensure early understanding of conformity requirements for different countries and regions.
- Involve relevant certification bodies early in the project.
  One common method to gain market access is to comply to relevant standards.

In the EU, they have different approaches for the different domains. For example, within the railway domain the safety standards are mandatory. A mandatory standard is, according to ISO/IEC Guide 2:2004, "Standardization and related activities General vocabulary": a standard the application of which is made compulsory by virtue of a general law or exclusive reference in a regulation.

Regulation (EU) No 1025/2012 "on European standardization" provides definitions for the terms "standard," "national standard," "European standard," "harmonized standard," and "international standard."

- "Standards" are defined as technical specifications adopted by a recognized standardization body for repeated or continuous application, with which compliance is not compulsory.
- "European standards" are "standards" adopted by European standardization organizations (ESOs).

  A "harmonized standard" means a European standard adopted on the basis of a request made by the Commission for the application of Union harmonization legislation.

Certification is intentionally not part of safety standards. Whether certification is necessary depends on the domain (e.g., need a NoBo—Notified Body—in EU), region/country, and the market.

Accreditation is a means of assessing, in the public interest, the technical competence and integrity of conformity assessment bodies. The idea of regulating accreditation at European level is twofold:

- A comprehensive European framework for accreditation provides the last level of public control in the European conformity assessment chain and is therefore an important element in ensuring product conformity.

- It enhances the free movement of products and services across the EU by underpinning trust in their safety and compliance with other issues of public interest protection.

The intention of the accreditation system is that the assessment and certification practices are acceptable worldwide or in regions like the EU, meaning that they are competent to, e.g., assess, test, and certify third parties. This is well illustrated by the IAF slogan "Certified Once Accepted Everywhere." This should also apply for safety assessments in the future.

Certified Once Accepted Everywhere

Several persons and companies that perform independent assessments have chosen or have been obliged by their customer or the national safety authority to be accredited for their services. To become accredited as an ISA, one has to comply with, e.g., one of the ISO 17000 standards—especially the ISO/IEC 17020:2012 "General criteria for the operation of various types of bodies performing inspection" has been used in Europe. ISO/IEC 17020:2012 specifies requirements for the competence of bodies performing inspection and for the impartiality and consistency of their inspection activities.

The evidence for an accredited certification is that the certificate and corresponding reports must have an accreditation mark on it. The accreditation mark should be from an accreditation body that have signed the MLA (Multilateral Recognition) for the technical activity you require (testing, measurement, verification, certification, etc.). It is possible to check whether the accreditation body is a signatory for the scope by checking the EA (European Accreditation) website at www.european-accreditation.org/ and for the world, International Accreditation Forum (IAF).

Copy from www.iaf.nu/International Accreditation Forum (copied 2021-06-25): "The IAF is the world association of Conformity Assessment Accreditation Bodies and other bodies interested in conformity assessment in the fields of management systems, products, services, personnel and other similar programmes of conformity assessment. Its primary function is to develop a single worldwide conformity assessment program that reduces risk for business and its customers by assuring them that accredited certificates may be relied upon. Accreditation assures users of the competence and impartiality of the body accredited."

A checklist that can be used to ensure that the ISA/CB is accredited is presented below.

1. Check that there is an accreditation mark on the ISA report.
2. Check that the accreditation mark is a mark of an accreditation body signatory to, e.g., the EA, ILAC (International Laboratory Accreditation Cooperation), or IAF (International Accreditation Forum) MLA (Multilateral Recognition Agreement). The marks can be seen at www.iaf.nu.
3. Check that your ISA is accredited for the competence, the tests, the results you need.

4. Check that the assessment has been carried out against European/international standards. Alternatively, check that the standards and methods used can be accepted in the country of destination.
5. In case of problems related to accreditation issues, contact the national accreditation body.

1.8   Prescriptive vs. Goal-Oriented Standards

1.8.1   Introduction

All the standards referred to in this book are prescriptive—at least to a certain degree. They require you first to do a risk assessment based on a set of parameters, e.g., the probability of an unwanted event, the event's consequences, and the possibility to avoid it. This will give you a risk level, such as SIL (IEC 61508) or an ASIL (ISO 26262). Given the risk level, the standard defines the required process activities. Some standards also state the requirements for the personnel involved, the tools used, and the resources allocated, e.g., IEC 61508-1:2010, which states that "Procedures shall be developed to ensure that all persons with responsibilities defined in accordance with 6.2.1 and 6.2.3 shall have the appropriate competence relevant to the specific duties that they have to perform."

Safety several standards will allow any life cycle process provided it meets certain requirements. See for instance IEC 61508-3:2010: "Any software lifecycle model may be used provided all the objectives and requirements of this clause (clause 7) are met." However, the main problem with using prescriptive standards is that they will lead or persuade you to stick to the mentioned techniques and methods related to the identified risk level, even after common practice has replaced them with something new and better. This problem is made more serious by the fact that it might be several years between each new version of a standard and we might thus be stuck with old methods and processes for a long time. The goal-oriented approach to standards solves this problem. Instead of telling you what to do, a goal-oriented standard will tell you what to achieve. Thus, the standard will be independent of changes in technology—methods and techniques. This will give us the following benefits: We can

- Choose whatever approach we want as long as we reach the defined goals.
- Start using new methods as soon as they become available and we have done the necessary testing, validation, and evaluation.

However, we need to be aware that there are several types of software development companies. At one extreme, we find companies that have developed safety-critical software for a long time and at the other extreme we find companies that are entirely new to the idea.

Well-experienced software development organizations are usually focused on the goals but want to become more resilient. For them, a standard such as IEC 61508 is a straightjacket even though clause C.1.3 "Method of use" of IEC 61508-3 states that: "Although Annex A recommends specific techniques, it is also permitted to apply other measures and techniques, providing that the requirements and objectives of the lifecycle phase have been met." A similar approach is suggested by ISO 26262-2: "It is allowed to substitute a highly recommended or recommended method by others not listed in the table, in this case, a rationale shall be given describing why these comply with the corresponding requirement."

For experienced companies, a recurring problem is that most of the standards force them to focus too much on the techniques and measures considered as best practices when the standard was last updated. However, for companies with little or no experience in the development of safety-critical software, the standard's process activity requirements are helpful since they tell what to do. Thus, goal-based standards are not for everyone. Some standards, e.g., ISO 26262:2018, have moved in a goal-based direction. As an example, this standard states the following: "When claiming compliance with ISO 26262, each requirement shall be complied with, unless one of the following applies:

- Tailoring of the safety activities in accordance with ISO 26262-2 has been planned and shows that the requirement does not apply.
- A rationale is available that the non-compliance is acceptable, and the rationale has been assessed in accordance with ISO 26262-2."

1.8.2   A Path to a Goal-Oriented Approach

In the list below, we have presented suggested future requirements to make the standards goal based. A goal-based approach allows us to use whatever tools, techniques, and methods are appropriate as long as we meet the standard's designated goals, the self-assessment is fulfilled, and they are accepted by the assessor:

- A safety life cycle for the development of the software shall be selected and specified during safety planning.
- Any software life cycle model may be used, provided all the objectives and requirements are met. Note that some standards already allow this—e.g., ISO 26262.
- Each phase of the software safety life cycle shall be divided into elementary activities with the scope, inputs, and outputs specified for each phase.
- Provided the software safety life cycle satisfies the life cycle requirements, it is acceptable to tailor the model chosen (e.g., V-model or SafeScrum) to take account of the safety integrity and the complexity of the project.
- Any customization of the software safety life cycle shall be justified on the basis of functional safety.
- Quality and safety assurance procedures shall be integrated into safety life cycle activities.

- For each life cycle phase, appropriate techniques and measures shall be used. Annexes A (EN 50128:2011 and IEC 61508-3:2010) and B (IEC 61508-3:2010) provide a guide.

The maritime industry, which has been using a goal-oriented approach for some time now, has complained that the goal-oriented standards give them much more work to do. Their argument is as follows: with a prescriptive standard they know what to do—just follow the list of activities. With the goal-based approach, the company has to come up with the tools and techniques needed to achieve the goals and convince the assessor and themselves that the choice will allow them to meet the specified goals. This requires high competence both from the development company and from the assessor. However, all the software development requirements in IEC 61508-3 are techniques and measures used to realize a set of sound software engineering practices.

1.8.3   The IMO 5-Tier Approach

The full 5-tier IMO model is shown below—see Huss (2007). International Maritime Organization (IMO) has later also included the need for a safety case together with the goal-oriented approach. According to Skjong (2005): "In 1992, the UK House of Lords Select Committee on Science and Technology recommended a Safety Case Regime for shipping, similar to that adopted in the UK oil and gas industries a couple of years earlier. It also recommended a move towards performance (or goal based) standards in place of prescriptive rules, and a concentration on the management of safety." IMO have also decided to have a goal-based approach for Maritime Autonomous Surface Ships (MASS). In 2021, they decided that the best way forward was to introduce MASS into the regulatory framework. The overall goal is clear: MASS should be as safe as conventional ships (Fig. 1.7).

The model is easier to understand if we focus on the three main parts, as shown in Fig. 1.8. It is important to be aware of the following:

- The development company cannot hold the assessor hostage to a decision—i.e., they cannot ask the assessor to specify how a job should be done.
- The assessor is not required to tell the development company why a suggested approach is not accepted even though they are free to do so. This might pose a challenge to the development company if their first approach is rejected, and they need to come up with a new one.
- If the company chooses a certain set of techniques and measures to meet a goal, the assessor has no responsibility for the outcome of the development.
- The development company should provide a safety case for the system developed.

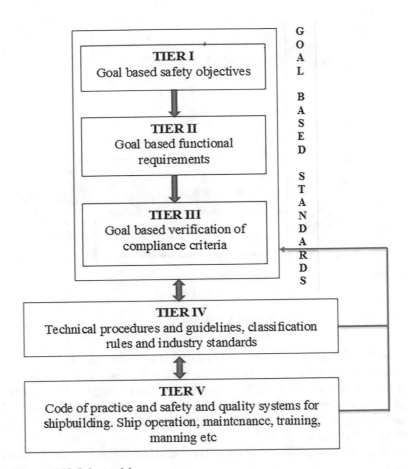

Fig. 1.7 The IMO 5-tier model

The process suggested for software development by this model is simple. The new, goal-based standard says what you should achieve, e.g., sufficient test, verifications, and analyses of all safety-critical components for this risk level. The development company makes a safety plan for how to achieve this goal and present it to the assessor who will either accept or reject the proposed plan—see also the Safety Plan chapter (Sect. 3.1) in this book. If it is rejected, the development company has to update the plan, hopefully based on the assessor's feedback. If the plan is accepted, the development company works according to it and make sure that it leaves a "paper trail" which can be used later to check if they have worked as previously agreed.

Fig. 1.8 Simplified IMO use of goal-oriented standards

1.8.4   The Need for a Safety Case

When we have used a goal-based approach during development, the confidence in the system's safety no longer rests with the standard(s) used. A goal-based safety proof/safety case is the natural approach to choose. The safety plan describes how it should be developed. An example of a safety case suggested for a goal-based safety management is shown in Fig. 1.9. Note that even here there are references to "appropriate industry standards."

The safety case suggested by Becht (2011) has three subgoals: we need to prove that:

- The safety requirements are complete.
- The system satisfies all explicit safety requirements. For example, if there is a requirement saying that the system should still be operating fine if it loses connection with a sensor, then there should be a claim stating that "the system will still be operating if it loses a sensor" and the necessary arguments and proofs should support the claim.
- The system is developed according to appropriate standards. When using goal-oriented development, this implies that we show that we have done what was agreed with the assessor.

A concept of operations—top-level environment for the safety case—is a document describing the characteristics of a proposed system from the viewpoint of an individual who will use that system.

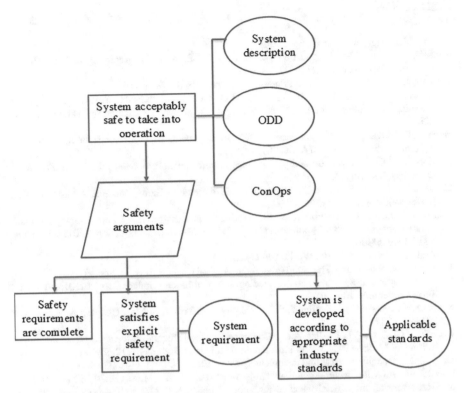

Fig. 1.9 Upper levels of an adapted figure according to Becht's suggested safety case of goal-based safety management

1.8.5 The Link to Safety Plan and the Safety Case

A safety case uses information created by the project to show that the system is safe. Thus, the safety plan and the safety case are tightly intertwined—the safety plan describes what the project needs to do to get the information that the safety case needs. The safety case needs three categories of information—information that is:

- Collected during the regular software development process—e.g., input to and results from a test
- Available in the company's databases—e.g., knowledge and experiences of each developer.
- Needed for the safety case only—e.g., proofs that there was a good communication with the customer or customer's representative

If we use an agile development process, it is important to get the activities needed to get the necessary information into the product backlog.

References

AAMI: Main page. www.aami.org (2020)

Ambler, S.: Agile Modelling: Effective Practices for Extreme Programming and the Unified Process. Wiley, New York (2002)

Ambler, S.: Ambysoft 2008 Agile Principles and Practices Survey (2008)

Anderson, W., Morris, E., Smith, D., Ward, M.C.: COTS and Reusable Software Management Planning: A Template for Life-Cycle Management. Software Engineering Institute, October 2007

Becht, H.: Moving towards goal-based safety management. In Proceedings of the Australian System Safety Conference (ASSC 2011) (2011)

Brodbeck, F.: Communication and performance in software development projects. Eur. J. Work Orgn. Psychol. **10**(1) (2001)

Cockburn, A.: Agile Software Development: The Cooperative Game, 2nd edn. Addison-Wesley, Upper Saddle River, NJ (2008)

eCampusOntario: Communication for business professionals. Open Library. https://ecampusontario.pressbooks.pub/commbusprofcdn/front-matter/introduction/ (2018). Last visited 7 May 2020

Harford, T.: How to Make the World Add Up (2021)

Horn, J.: Marx Was Right! The Guilds and Technological Change. J. West. Soc. Fr. Hist. **33** (2005)

Huss, M.: Status at IMO: Where are we heading with goal-based standards? SAFEDOR—The Mid Term Conference, May 2007.

Kniberg, H., Ivarsson, A.: Scaling Agile @ Spotify with Tribes, Squads, Chapters & Guilds (Oct 2012)

Kroeck, K.G., Kirs, P.J., Fiedler, A.M.: Cognitive Biasing Effects in Information Systems: Implications for Linking Real World Information with Human Judgment (1989)

Levy, D.M.: Document reuse and document systems. Electron. Publ. **6**(4), 339–348 (1993)

Lucidchart.: www.lucidchart.com/blog/what-is-a-tiger-team (2020). Last visited 18 May 2021

MISRA: Website, the official resource for information on MISRA's publications and activities. www.misra.org.uk (2021). Last visited 27 April 2021

Myklebust, T., Stålhane, T.: TrustMe, we have a safety case for the public. ESREL (2021)

Myklebust, T., Stålhane, T., Hanssen, G.K., Haugset, B.: Change impact analysis as required by safety standards, what to do? PSAM12, Hawaii, 2014

NewsCred: How our Engineering Team Uses Guilds to Increase Collaboration (2021)

Skjong, R.: Formal safety assessment and goal based regulations at IMO—Lessons learned. In: Proceedings of OMAE2005, 24th International Conference on Offshore Mechanics and Arctic Engineering (OMAE 2005), Halkidiki, 12–17 June, 2005

Šmite, D., Moe, N.B., Levinta, G., Floryan, M.: Guilds: How to succeed with knowledge sharing in large-scale agile organizations. IEEE Softw. **36**(2), 51–57 (2019)

Stålhane, T., et al.: Hva ønsker programvareindustrien av våre kandidater? In: Norwegian, The MNT Conference, Tromsø, 2019

Taipuva.: www.taipuva.com/articles/reuse-requirements-standards-efficiently-projects/ (2018)

Wang, Y., Graziotin, D., Kriso, S., Wagner, S.: Communication channels in safety analysis: An industrial exploratory case study. J. Syst. Softw. **153**, 135–151 (2018)

Wikipedia: Group thinking (2021)

Wynekoop, J.L., Walz, D.B.: Characteristics of high performing IT personnel: a comparison of IT versus end-user perception. In: Agarwal, R., Prasad, J. (eds.) Proceedings of the ACM SIGCPR Conference, April, New Orleans (1999)

Chapter 2
Agile Practices

The world has a way of undermining complex plans. This is particularly true in fast moving environments. A fast moving environment can evolve more quickly than a complex plan can be adapted to it. By the time you have adapted, the target has changed.

Carl von Clausewitz—Vom Kriege

What This Chapter Is About
- Benefits and reasons for going agile.
- Some information on agile practices.
- Challenges and their agile solutions.
- An alternative view.
- Requirements management.
- On alongside engineering.
- Some practices are extended to accommodate safety aspects.

2.1 Overview

The main reasons for including practices and their benefits are summed up in Figs. 2.1 and 2.2. Some of the practices collected under the heading "Agile Practices" are specific to agile methods—e.g., the daily scrum—while some are old—e.g., incremental development (Larman and Basili 2003). In addition, some of them are old but are used in a new way—e.g., post-mortem analysis (Birk et al. 2002), now called retrospectives. According to Vardhman (2020), 70% of all software development companies use one or more agile development practices. We will discuss three aspects of agile practices: (1) which practices are often used (VersionOne 2020), (2) which have been selected most often in the development of safety-critical software (Doss and Kelly 2016), and (3) their adaptations when applied to the development of safety-critical systems.

© The Author(s), under exclusive license to Springer Nature Switzerland AG 2021
T. Myklebust, T. Stålhane, *Functional Safety and Proof of Compliance*,
https://doi.org/10.1007/978-3-030-86152-0_2

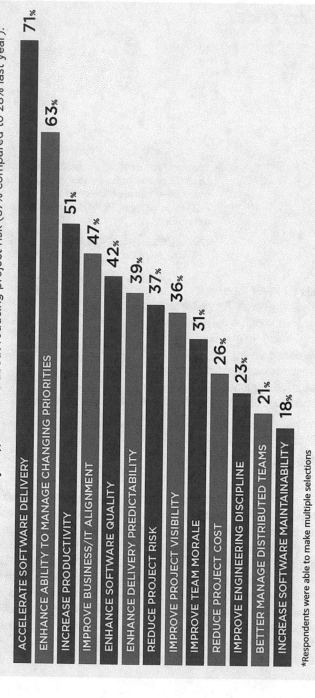

Fig. 2.1 Reasons for adopting agile practices. Copy from https://stateofagile.com/# © Digital.ai

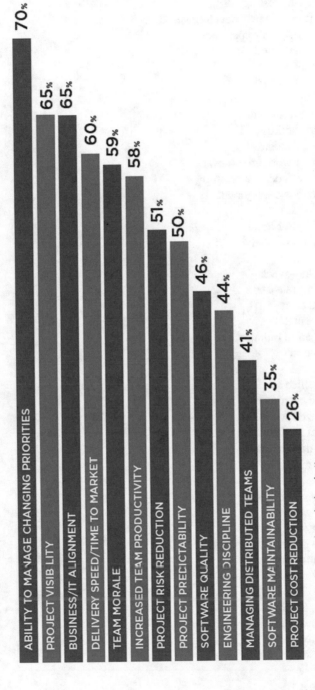

Fig. 2.2 Benefits from acopting agile practices. Copy from https://stateofagile.com/# © Digital.ai

55 Agile practices

1. Acceptance testing
2. AFD (analysis first development (new))
3. Agile UX (user experience)
4. Automated build
5. Automated tests
6. Burn down chart
7. Backlog
8. Backlog refinement
9. Backlog splitting
10. Coding standard
11. Common work area (new)
12. Collective code ownership
13. Continuous deployment
14. Daily scrum
15. Definition of done
16. Definition of ready
17. Epic
18. 40-h work week
19. Frequent releases
20. Hazard story
21. Incremental
22. Information radiators
23. Integration
24. Iteration
25. Open office space
26. On-site customer
27. Prioritized work list
28. Product road mapping
29. Quick design session
30. Refactoring
31. Release plan
32. Relative estimation
33. Safety story
34. Security story
35. Simple design
36. Shippable code
37. Short iterations
38. Sprint
39. Sprint planning
40. Sprint retrospective
41. Sprint review
42. Stepwise integration
43. Story
44. Task board
45. Team-based estimation

46. Team communication
47. Test-driven requirement
48. Test first development
49. The wall
50. Time box
51. Unit testing
52. Usability testing
53. Velocity
54. Version control
55. BDD (behavior-driven development)

Recently, we have seen an increasing use of agile practices when developing safety-critical software (SCSW) to reduce time to market, reduce costs, and improve quality. The survey performed by Myklebust et al. (2017a, b) shows that even the conservative railway signalling industry has started experimenting with agile methods. Companies introducing agile methods like Scrum also have to include relevant Agile practices to get the full benefit of an agile approach. A practice in software development is a working activity (e.g., writing a code, putting yellow stickers with text on a board), and it is required that the activity can be repeated. We have limited the practices to agile practices, but some may also be used when following a waterfall or V-model approach.

For safety-critical systems, however, some agile development practices do not fit as is but have to be adapted and/or extended to accommodate a safety approach. Below, the agile approach and agile practices are described, both in general terms and—more specifically—how some of them are used as part of SafeScrum (Hanssen et al. 2018). Unfortunately, little experience has been published on using agile practices for use together with safety standards and safety-critical software in general (Table 2.1).

Table 2.1 Most used agile practices

11th 2017	12th 2018	13th 2019	14th 2020
1. Iteration planning	1. Daily stand-up	1. Daily stand-up	1. Daily stand-up
2. Daily stand-up	2. Sprint/iteration planning	2. Sprint/iteration planning	2. Retrospective
3. Retrospectives	3. Retrospectives	3. Retrospectives	3. Sprint/iteration planning
4. Iteration reviews	4. Sprint/iteration review	4. Sprint/iteration review	4. Sprint/iteration review
5. Short iterations	5. Short iterations	5. Short iterations	5. Short iterations
6. Release planning	6. Release planning	6. Planning poker/team estimation	6. Kanban
7. Team-based estimation	7. Planning poker/team estimation	7. Kanban	7. Planning poker/team estimation
8. Dedicated product owner	8. Kanban	8. Release planning	8. Dedicated customer/product owner
9. Single team	9. Dedicated customer/product owner	9. Dedicated customer/product owner	9. Release planning
10. Frequent releases	10. Single team	10. Single team	10. Single team (integrated development and test)

There exist more than 80 named agile practices, as shown by, e.g., the Agile Alliance and VersionOne (2020), but we will only look at a few of them. Several of these practices cannot be used as is when developing SCSW as they do not include parts that are mandatory for safety-critical development. We have evaluated 20 (Table 2.2 and Sect. 2.2) of the most relevant practices. Necessary add-ons and adaptions are described in Sect. 2.2 to ensure that important international standards such as IEC 61508:2010, ISO 26262:2018, and EN 50128:2011 can be satisfied.

Unfortunately, few papers published discuss the important question of which are the most important agile practices. Some of the following information is taken from blogs. This is not surprising—blogs are mostly 2–4 years ahead of scientific papers when it comes to the cutting edge for most aspects of software engineering.

The VersionOne (2020) report shows that the five most used agile practices are the daily stand-up/daily Scrum (85%), retrospectives (81%), sprint reviews (77%), short iterations (64%), and Kanban (69%), while, for instance, only 24% use Agile/ Lean UX (user experience)—the "think-make-check" loop—similar to the *Lean Startup* method's "build-measure-learn" feedback loop (Ries 2011). Two-thirds of the survey respondents said they worked in software organizations with more than 100 people. About 50% of the respondents worked for technology firms, financial services, and professional services. A total of 7% of the industrial respondents using agile development was safety related—4% from healthcare and 3% from transportation. In addition, 41% of the respondents are from North America and 31% from Europe.

T. Kelly has performed a survey in the UK. Among the questions used in the survey, one is of special interest here: "Which of the following practices of agile development can contribute to safety-critical systems development and assurance?". They had 69 respondents; thus, all differences larger than 0.12 are significant at the 5% level. The five most popular results were as follows:

1. Simple design—67%
2. Continuous integration—61%
3. Release planning—56%
4. Pair programming—50%
5. Small and short releases—44%

Except for "short releases," the two top five sets are disjoint. The large differences between the survey results from VersionOne (2020) and Doss and Kelly (2016) can, at least partly, be explained by different survey populations. VersionOne (2020) looks at the whole software community, while Kelly looks only at those developing safety-critical software. The top five practices from VersionOne are mostly related to administrative matters, while three of Kelly's top five items are related to development—e.g., simple design. Thus, it seems that what are important agile practices will depend on the area of interest. In reality, one has to handle both administrative and software development concerns; thus, both areas are important.

However, the Chaos report identifies that the two main reasons for projects to go wrong are bad management and bad communication. The Chaos report has received a certain amount of complaints from academia—see, e.g., Eveleens and Verhoef

Table 2.2 Engineering practices

Engineering practices	Explanations and comments
1. Unit testing	We will focus on test first development (TFD)—mainly because of its popularity as it enables good code design and code documentation. TFD is a development practice that embraces the principle of never adding or changing code without first having added or changed the runnable test case that verifies the code's success criteria. The test of the code—usually a unit-test——is defined before the code itself is developed. By constantly focusing on building tests prior to code, we will gradually grow up-to-date tests that cover the complete system. One of the benefits of using TFD is that the software is written in smaller units that are less complex and thus more testable, because more consideration is given to design issues

2. Coding standards	Required to be used by safety standards. A programming language coding standard shall:
	  1. Specify good programming practice
	  2. Proscribe unsafe language features—constructions that should not be allowed or only allowed under specific, documented circumstances
	  3. Promote code understandability. This is important for code reviews, for example
	  4. Facilitate verification and testing
	  5. Specify procedures for source code documentation
	Note that even though there is a coding standard for C++, the language should be used with caution for safety-critical systems. C++ is in the position once held by "C"; many people believe that it should not be used for critical systems, but its use within the field is growing, and that growth is without a common set of guidelines. See also Misra

3. Refactoring	The containment of refactoring is important when complying safety standards because an apparently minor modification to one section of source code could have major impact on requirements documents, design documents, requirements-based tests, or systems tests
4. Automated acceptance testing	See Sect. 2.1.2
5. TDD/TFD/AFD	TDD: Test-driven development has especially one component that is important, that is, the test first approach (TFD) AFD (Myklebust et al. 2019): Analysis first development (AFD) could be seen as an approach similar to test first development (TFD). TFD refers to programming activities in which three activities are tightly connected: coding, testing (in the form of writing unit tests), and improvements (in the form of, e.g., refactoring). Our suggested AFD approach includes safety analysis before the first sprint, during the first sprint (iteration), as a part of alongside engineering (performed by experts not part of the software sprint team, e.g., safety experts), and as part of change impact analysis
6. BDD	See Sect. 2.1.2
7. Definition of ready	See section below
8. Definition of done	See section below
9. Safety stories	(Myklebust and Stålhane 2016). Safety stories are a fairly new practice, developed to ensure that the agile requirements management process encompasses the important safety requirements and required measurements and techniques. The safety story card will create a common

(continued)

Table 2.2 (continued)

Engineering practices	Explanations and comments
	problem understanding between the software developers and the safety stakeholders. Safety stories are user stories that include one or more safety requirements
10. Simple design	Simple design bases its software design strategy on the following principles: – Design is an ongoing activity, which includes refactoring and heuristics such as YAGNI ("You Aren't Gonna Need It"). This has similarities with MVP (minimum viable product) Design quality is evaluated based on the rules of code simplicity and design elements such as "design patterns," plugin-based architectures, or similar. Design decisions should be deferred until the "last responsible moment." Therefore, it is important to collect as much information as possible on the benefits of the chosen option before taking the costs Safety adaptations: Safety-critical systems include both functional and safety-critical requirements. That makes them complex. As a result, it is even more important to have simple design in mind. Regarding "last responsible moment," this should be discussed with the safety experts

(2010). Their main objection is the definition of two of the project categories—success and challenge—based on adherence to time and budget. In this book, however, only the category "impaired" is used. In addition, the IT professional facilitator (Standish 2009) has published their top ten list of causes for failure. They are all related to management and communication. However, just focusing on avoiding failure is not sufficient. We also need to develop a product, and we thus add development practices. This leads us to recommend projects to focus on the following agile practices, which are related to the following three aspects:

- Communication—how is the information flow organized in the project? The number one priority for any successful project.
- Planning—what to do when.
- Development—how to do it.

In the subchapters below, the most important agile practices are evaluated for SCSW. In addition, we describe extensions of two agile practices.

2.1.1 Practices Suitable Both for Traditional and Agile Processes

We have evaluated which practices can be used both when having a traditional and an agile approach. The evaluation has been based on the author(s) experience with companies using these practices also when having a traditional development process (Table 2.3).

Table 2.3 Practices that can be used both in traditional and agile projects (alphabetical order)

Practice that can be used both in traditional and agile projects	
1. Acceptance testing	8. Prioritized worklist
2. Automated tests	9. Refactoring and simple design
3. Backlog (backlog splitting, backlog refinement, and backlog grooming)	10. Stories (user stories, safety stories, and hazard stories)
4. Daily scrum/four questions	11. Test first development
5. Definition of done and definition of ready (FAA 2000), ISO/IEC/IEEE 26515:2018, and AAMI (2012)	12. Usability testing
6. Epic	13. Version control
7. Analysis first development (Myklebust et al. 2019)	14. Coding standard

2.1.2 Practices Suitable in the First Safety Phases

We have evaluated which practices are relevant for the first safety phases as part of the work described above. Upfront activities are also important when having an agile approach (Table 2.4).

Table 2.4 Practices relevant in the first safety phases

Before the first sprint	The first nine safety phases according to IEC 61508
1. Iteration planning 2. Release planning 3. Backlog 4. Backlog splitting 5. Backlog refinement 6. Coding standard 7. Definition of ready 8. Data readiness 9. Epic 10. Prioritized work list 11. Relative estimation 12. Team-based estimation 13. Story 14. Story mapping 15. User story 16. Safety story 17. Hazard story 18. Security story 19. Task board 20. Test-driven requirement 21. The wall	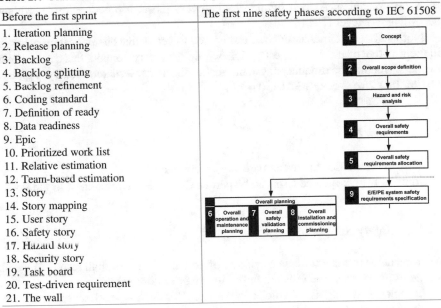

2.1.2.1   Popular Practices Relevant for SCSW

The top ten Agile software practices were established as a combination of the most frequently used practices and our evaluation of how the practices fit into the development of safety-critical software. The four Scrum practices, sometimes also named ceremonies, are included together with six other practices.

2.2   More Detailed Information for Some of the Agile Practices

2.2.1   Acceptance Testing: A Development Practice

2.2.1.1   Information

An acceptance test is a formal description of the behavior of a software product, generally expressed as an example of a usage scenario.

An acceptance test generally has a binary result, pass or fail.

Acceptance testing has some similarities with TFD (test first development). Required acceptance tests should be planned for and developed as early as possible. TFD is a good practice, and parts of the acceptance tests should be planned together with TFD planning. Associating a test with every piece of functionality is brilliant according to Meyer (2014).

Normally, a customer acceptance test is used to verify that an application behaves in the way that a customer expects, while within the safety domain, this can be a test to check whether the product or system satisfies the safety tests required by, e.g., EN 50128, ISO 26262-6, or IEC 61508-3.

2.2.1.2   Benefits

- Early evidence that the tested parts are acceptable.
- Some of the acceptance tests can be part of the evidence referenced in the PoC.

2.2.1.3   Safety Adaptations

An acceptance test is a formal description of how the software shall respond related to one or more tests. The tests are linked to safety requirements.

Acceptance testing is of crucial importance for SCSW, and several tests are required as part of the verification and validation plan. Regression testing is especially important when using incremental SW development. This topic should be improved in the next edition of both EN 50128 and IEC 61508-3 and should also be improved in ISO 26262. Some of the acceptance tests will be performed by the independent testers in the safety or alongside engineering team (see Sect. 3.2) during a sprint. The final tests are run after the end of the last Sprint if, e.g., coding has been performed as part of the last Sprint.

2.2.2   Backlog, Backlog Splitting, and Backlog Refinement: Development and Planning Practice

2.2.2.1   Agile Definition

Backlog: Defined in ISO/IEC/IEEE 26515:2018 as "3.4 backlog: *collection of agile features or stories of both functional and non-functional requirements that are typically sorted in an order based on value priority*."

- Backlog splitting (Stålhane et al. 2012) includes tagging or separating the safety requirements.
- Backlog refinement consists of reviewing items in the backlog to ensure the backlog contains the appropriate items, that they are prioritized, and that the items at the top of the backlog are ready for delivery. This activity occurs regularly and may be an officially scheduled meeting or an ongoing activity. Some of the activities that occur during this refinement of the backlog include:

  - Removing user stories that no longer appear relevant
  - Creating new user stories in response to newly discovered needs
  - Reassessing the relative priority of stories
  - Assigning estimates to stories which have yet to receive one
  - Correcting estimates in light of newly discovered information
  - Splitting user stories which are high priority but too coarse grained to fit in an upcoming iteration

- Scrum product backlog is simply a list of all things that need to be done within the project. It replaces the traditional requirements specification artifacts. The Scrum team then determines which items they can complete during the coming sprint (Table 2.5).

2.2.2.2   Benefits

- Auditable safety requirements
- Easy for the development engineers to know the safety requirements
- Improved link between the functional and non-functional requirements
- Improved link between the SRS, the backlog, and the hazard log
- Improved understanding of the requirements

Table 2.5 Example backlog

Backlog			
Item	Safety tag	Size	Sprint no.
Safety requirement 1	*	2	1
Safety requirement 2	*	5	
Functional requirement 2		10	
Safety requirement 2	*	12	2
Functional requirement 3		8	
Functional requirement 4		6	3
Safety requirement 3	*	14	
Functional requirement 5		9	4
Functional requirement 6		8	

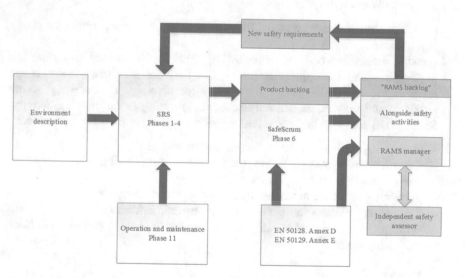

Fig. 2.3 SafeScrum and separation of responsibility between the Sprint team and the alongside engineering when applying the railway safety standards. Extension based on Fig. 4.1 in Hanssen et al. (2018)

2.2.2.3   Information and Safety Adaptations

- Backlog refinement: Similar to the agile approach but more focus on hazard and safety stories together with the understanding of them and the link to the user stories. The safety manager or RAMS manager should be included in these meetings.
- Evaluate the link between hazards, SRS, and hazard/safety stories. The manufacturer may also perform this as part of the RHA (requirement hazard analysis); see Sect. 4.5.3.
- A prioritized work list is also closely linked to this practice (Fig. 2.3).

Backlog splitting is an important part of SafeScrum (Stålhane et al. 2012) and the Agile HL approach (Myklebust et al. 2017a, b). This practice was introduced as part of the introduction of SafeScrum.

In SafeScrum, the set of requirements is split into safety-critical requirements and other requirements and inserted into separate product backlogs. Alternatively, the safety requirements are tagged. Adding a second backlog is an extension of the original Scrum process and is needed to separate the frequently changed functional requirements from the more stable safety requirements. With two backlogs, we can track how each item in the functional product backlog relates to the items in the safety product backlog, i.e., which safety requirements are affected by which functional requirements. This can be done by using simple cross-references in the two backlogs or can be supported by adding an explanation of how the requirements are related if this is needed to understand a requirement. The staffing of the *Sprint team* and the duration of the sprint (1–4 weeks is common), together with the estimates of each item, decide which items can be selected for development. When matters related to safety are on the table, the RAMS responsible person (see Chap. 3) should also participate in selecting which items have to be prioritized.

It will be convenient to extend the safety stories with two pieces of information— how can you discover that something has gone wrong and how can you fix it? The idea of Leite et al. (2020) is to define two new types of stories—failure detection stories and failure containment stories. In our opinion, the important point here is not the story format but that the developers from the start of the project for each safety requirement consider (1) how can we discover that we enter an unsafe state and (2) what can we do about it? By doing this, we will increase safety awareness and thus the safety culture. The diagram below is an example taken from the oxygen saturation control (SPO2) in an Integrated Clinical Environment (ICE) (Fig. 2.4).

Backlog refinement is also known as story time and backlog grooming. The product backlog needs to be refined based on current knowledge. The team, the Scrum master, the product owner, and, e.g., the RAMS manager should participate in the backlog refinement meeting. Several of the safety requirements are taken care of mainly by the RAMS team. If in doubt regarding safety requirements related to legislation or standards, the assessor should be consulted. The backlog refinement meetings will improve the understanding of the requirements and as a result ensure that requirements are implemented correctly. In most cases, the backlog refinement process will not require SRS changers. However, if there are reasons to believe that the SRS should be changed, a CIA (change impact analysis) should be initiated.

2.2.3 Prioritized Work List: A Development Practice

2.2.3.1 Information

Prioritizing the backlog items may be performed by using index cards or similar or just discussing this as part of the sprint planning.

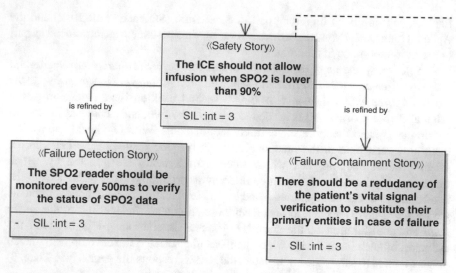

Fig. 2.4 Safety story refinement according to © Leite et al. (2020)

2.2.3.2   Benefits

- Important to ensure that the most important parts are prioritized.
- Ensure that the safety parts are considered early in the project.

2.2.3.3   Safety Adaptations

The RAMS manager or, e.g., the safety manager should be involved in the prioritization.

Together with the Agile safety plan (Myklebust et al. 2016a, b), the high-level safety plans and the Sprint planning constitute the main Agile plans.

This topic is an important part of SafeScrum (Stålhane et al. 2012) and the Agile HL approach (Myklebust et al. 2017a, b).

2.2.4   Daily Scrum Meeting (DSM) Including Four Questions: A Communication Practice

2.2.4.1   Information

Findings by Stray (2016) show that DSMs may not necessarily have to be held daily and focus in the meetings should be on discussing and solving problems, and planning for the future, rather than reporting what has been done. Furthermore, it is beneficial to be standing in the DSMs and to conduct the meetings by a task board. This empirical evidence corresponds to the experience by one of the authors.

2.2.4.2   Benefits

- Early problem-solving and corrections
- Short-term planning

2.2.4.3   Safety Adaptations

Paasivaara (2008) examined agile practices in global software development and found that DSMs helped reveal problems early, which is important when developing SCSW.

Regarding daily scrum and four questions (Myklebust et al. 2016b), see the next chapter.

The four questions consist of the three questions usually used in the daily Scrum plus one extra question related to safety.

1. What work did you complete yesterday?
2. What have you planned for today?
3. Are you facing any problems or issues?
4. Any safety-related impact of the completed work?

The fourth question in the list is especially important for organizations that develop both safety and non-safety products. It is relevant both for the work performed yesterday and for the work to be performed today. In addition, it helps to build a safety culture in the team and later also in the company.

If the answer to question 4 is positive, we need some additional process activities. First, we need to close the daily stand-up meeting. Those who have the necessary competence stay for the safety meeting to discuss and resolve the safety issues. If this proves difficult, we should involve a safety expert, and, if this also fails, we should involve the assessor. Sometimes the answer may be "I do not know" or "I'm not sure." This should be followed up by, e.g., the RAMS manager together with a person from the Sprint team.

There is no need to record any meeting minutes; the value of the daily stand-up is to keep everybody informed and quickly highlight any problems. The Scrum master is responsible for taking actions if there are any problems. Detailed discussions should be avoided in the meeting in order to keep it short and informative. If there is a need to discuss specific details, this should be done after the meeting and only involve those needed—leaving others to continue their work (Fig. 2.5).

2.2.5   Sprint Planning Meeting: A Planning Practice

2.2.5.1   Information

The sprint planning meetings are attended by the product owner, the Scrum master, and the entire Scrum team. According to Kniberg (2015), sprint planning is a critical

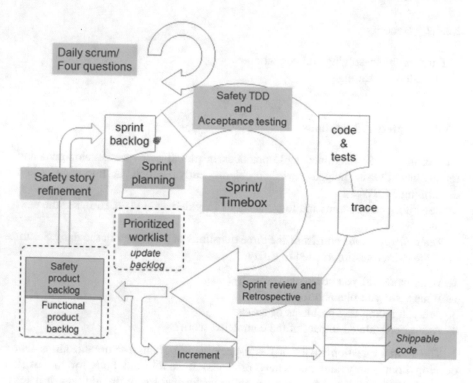

Fig. 2.5 Sprint together with relevant Agile practices. Extension based on Fig. 2.5 in Myklebust and Stålhane (2018)

meeting, maybe the most important event in Scrum. The sprint is the basic building block of Scrum, and the sprint planning meeting decides what shall be done in the next sprint.

2.2.5.2   Benefits

Planning is important even in an agile setting and includes among other things prioritization.

2.2.5.3   Safety Adaptations

The planning has to take into account safety aspects. Maybe, e.g., it is of crucial importance to do some of the safety tests earlier.

2.2.6   *Time Boxing: A Development and Management Practice*

2.2.6.1   Information

The sprint should have a fixed length and is time boxed—i.e., has a fixed length ranging from 2 weeks to a month. Scrum meetings last about 10–15 min each day.

2.2.6.2   Benefits

According to the Spica blog (Spica), time boxing is good because you can:

- More easily "force yourself" to start working on the tasks you procrastinated on or you know they're hard for you to be doing.
- More easily set strict limits on how much time you'll spend on a specific task and when you will spend it, and so you can organize yourself much better.
- Boost your productivity and focus greatly if you make sure that nobody interrupts you or distracts you while you're working on your task in the boxed time for it.
- Use time boxing to plan the most important things right in the morning, and it can help you to plan a much better working rhythm in general.

Although the benefits listed above refer only to one person, the same benefits hold for a Sprint team.

2.2.6.3   Safety Adaptations

The time box is an important concept both for Scrum and for SafeScrum. It is important for the management to ensure that the project is time boxed and does not change the planned Sprint activities due to, e.g., other projects. See also the section on incremental development below.

2.2.7   *Incremental Development Including Iteration and Stepwise Integration: A Development Practice*

2.2.7.1   Information

Development of software is normally incremental, e.g., product releases and maintenance releases are normally issued in increments improving the product. Research at the Standish Group indicates that shorter timeframes for, e.g., one or more sprints, with delivery of software components early and often, will increase the success rate. Shorter sprint timeframes result in an iterative process of designing, prototyping, development, testing, and deployment of small elements (Chaos 1995).

According to Meyer (2014), iteration is brilliant: *Short iterations are perhaps the most visible influence of agile ideas, an influence that has already spread throughout*

*the industry. Few competent teams today satisfy themselves with 6-month objectives.
The industry has understood that constant feedback is essential, with a checkpoint
every few weeks.*

2.2.7.2   Benefits

- Learning by doing
- Most important features added

2.2.7.3   Safety Adaptations

A stepwise integration as part of sprints is an integrated part of the SafeScrum
approach. Often there are a few iteration sprints before an integration (increment) is
performed to create a testable system. The certification process is also comprehensive, so normally a few features and some improvements are added for each
increment including the certification of the product or system. The agile experts
often mention "continuous integration" instead of "stepwise integration," but "continuous integration" is more difficult when developing SCSW. For example, safety
cases may have to be finalized, and assessors may be involved.

Incremental development is an important part of SafeScrum (Stålhane et al. 2012)
and the Agile HL approach (Myklebust et al. 2017a, b). See release management—
release plans and release notes—in Sect. 8.3.

2.2.8   *Shippable Code: A Development Practice*

2.2.8.1   Information

The idea of "shippable code" is also named "delivering working software" by Meyer
(2014): *The emphasis on delivering working software is another important contri-
bution. We have seen that it can be detrimental if understood as excluding requirements, infrastructure, and upfront work. But once a project has established a sound
basis, the requirements to maintain a running version impose a productive discipline
on the team.*

2.2.8.2   Benefits

- Software that are working
- Discipline on the team

2.2.8.3   Safety Adaptations

Several opponents against Agile have claimed that this is difficult when developing SCSW, but our studies and experience have shown that this is possible. But the code is not "shipped" as often as when developing traditional software.

See also release notes in Sect. 8.3.

2.2.9   Sprint Review: A Communication Practice

2.2.9.1   Information

At the end of each sprint, a sprint review meeting is held. VersionOne (2020) has iteration reviews ranked as number 4.

2.2.9.2   Benefits

- Software that are working
- Discipline on the team

2.2.9.3   Safety Adaptations

The sprint review is an important part of SafeScrum and the Agile HL approach (Myklebust et al. 2017a, b).

In SafeScrum, after a planned number of sprints, the project is required to deliver a potentially reviewable product increment. This means that at the end of each sprint, the team has produced a coded, tested, and reviewable piece of software.

The RAMS manager or, e.g., the safety manager may be involved in the sprint review.

According to Doss and Kelly (2016), systematic (design) errors are introduced whenever there is a misalignment of the original intent of a requirement and its implementation.

Potentially hazardous emergent behaviors could firstly result from well-intended but in hindsight flawed design decisions made when addressing or satisfying requirements that, unfortunately, have unintended hazardous side effects. Secondly, they can also result from implementation (process execution) errors during the software development process—e.g., modelling errors, coding errors, and tool-use errors. Therefore, it is necessary to ensure that assurance effort has been targeted at attempting to reveal both of these sources of errors.

2.2.10 Retrospective: A Communication Practice

2.2.10.1 Information

This is a dedicated period at the end of one or more sprints to reflect on how they are doing and to find ways to improve.

Quoting Kniberg: "Sprint planning is a critical meeting, probably the most important event in Scrum (in my subjective opinion of course). A badly executed sprint planning meeting can mess up a whole sprint. Important? Yes. Most important event in Scrum? No! Retrospectives are waaay more important!" (Kniberg 2015). VersionOne (2020) has retrospectives ranked as number 2.

2.2.10.2 Benefits

- Improves communication
- Improves the process

2.2.10.3 Safety Adaptations

This is an important part of SafeScrum. The RAMS manager or, e.g., the safety manager may be involved in the retrospectives.

2.2.11 Definition of Ready: A Development Practice

2.2.11.1 Information

Definition of "ready" is a development practice. The definition usually refers to the readiness of all activities necessary to start developing software. The concept of "ready and done" is critical in safety-critical development projects. Because of the emphasis on delivering both usable and safe software and achieving results, defining what ready means is important. Variations and phrases include "definition of ready" and "readiness," as in the degree of achieving "ready," e.g., ready for complete and required safety tests.

Definition of ready is linked to several other practices:

- Sprint planning (Myklebust et al. 2016a, b). Definition of ready is an important part of this meeting. We have to conclude which parts are defined as ready.
- Acceptance testing (e.g., FAT and SAT); see Sect. 9.1.
- Test first development (Myklebust et al. 2019).
- Definition of done; see section below.

- User stories, safety stories, and hazard stories (Stålhane and Myklebust 2018). For example, when a story is done; see list below that should be considered:

A story can be marked as ready when the stories:

- Meet *all* the related acceptance criteria (functional and non-functional requirements)
- Have been peer-reviewed
- Are independent of all other stories
- Are valuable (e.g., a vertical slice gives direct value to the product, but a horizontal slice may be valuable at a later increment)
- Are estimable (to a decent approximation)
- Are small (fit in a single iteration)
- Are testable (even if the test doesn't exist yet)
- Are analyzed for safety and hazard
- Are accepted by the product owner and the RAMS manager
- Have at least one test case or an analysis associated

2.2.11.2   Vertical Slice and Horizontal Slice

What we call vertical slice is similar to the concept of MVP (minimum viable product) from lean development, where you create the most basic piece of software that gives some value to the user. But sometimes you may be planning your iterations making horizontal slices, grouping, e.g., functionalities by technical topics instead of functional ones.

Horizontal stories (vertical slice) may not always meet a well-formed story structure of "As a <role>, I want <functionality>, so that I can achieve <value/outcome>," and they don't always meet the INVEST (independent, negotiable, valuable, estimable, small, testable) criteria.

You may plan for a vertical slices when:

- You're adding to existing functionality in an incremental fashion
- There is strong alignment between your data and business logic
- You're implementing new features with few dependencies

You may need to focus on horizontal slices when:

- New feature requests require new ways of thinking about existing functionality
- The roadmap includes sharing functionality across previously siloed products
- You encounter scaling pain points (Fig. 2.6)

2.2.11.3   Definition

The definition is usually referring to readiness of all activities necessary to start developing software. This means that the stories have to be well written and understood by the development team. See list above.

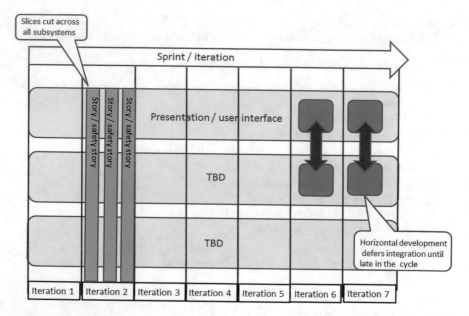

Fig. 2.6 Horizontal and vertical slicing

2.2.11.4 Benefits

- Does not start working on a feature or safety function that is poorly understood.
- Ensure that definition of done is achieved.
- Less technical debt.

2.2.11.5 Context

Definition of ready is related to safety and security domain requirements.

2.2.11.6 Safety Adaptations

- Safety and hazard stories must be analyzable
- The RAMS manager has accepted the safety and the hazard stories

2.2.12 Definition of Done: A Development Practice

2.2.12.1 Information

Definition of "done" is a development practice. The definition usually refers to doneness of all activities necessary to have shippable code. The concept of "ready and done" is critical in safety-critical development projects. Because of the emphasis on delivering both usable and safe software and achieving results, defining what done means is important. "Done" may have different meanings depending on whether this is part of, e.g., a product or system since integration tests may have to be performed later.

Definition of done is linked to several other practices.

- Sprint planning and the link to the agile safety plan (Myklebust et al. 2016a, b). Definition of done is an important part of this meeting. We have to conclude which parts are defined as done. This has also often to be discussed with, e.g., the RAMS manager.
- Acceptance testing; see Sect. 9.1.
- Test first development (Meyer 2014).
- Definition of ready; see Sect. 2.1.
- Shippable code; see Sect. 2.1.
- User stories, safety stories, and hazard stories (Myklebust and Stålhane 2016; Stålhane and Myklebust 2018). For example, when a story is ready, see list below of what should be considered:

A story can be marked as done when the story:

- Meets *all* acceptance criteria (functional and non-functional requirements)
- Has been peer-reviewed, e.g., by the RAMS manager for the safety stories
- Is independent of all other stories
- Is estimable (to a decent approximation)
- Is small (fit in a single sprint/iteration)
- Is testable (even if the tests don't exist yet)
- Is analyzed for safety and hazard
- Is accepted by the product owner and the RAMS manager
- Has at least one test case or an analysis associated with each requirement

2.2.12.2   Definition

The definition of "done" usually refers to doneness of all activities needed to have a shippable code with a corresponding safety case that is accepted by, e.g., a safety assessment report issued by an assessor. This means that the stories have to be well written and understood by the development team. See list above.

2.2.12.3   Benefits

- Prevent developers from starting to work on a feature or safety function that is poorly understood.
- Ensure that a shippable code is delivered, including the relevant evidence.
- Ensure that the relevant information has been developed and accepted.

2.2.12.4 Context

Definition of done is related to safety and security domain requirements.

2.2.12.5 Safety Adaptations

- Safety and hazard stories must be analyzable.
- The RAMS manager has accepted the safety and the hazard stories.
- In several domains, the shippable code is not done before the safety cases have been issued and are accepted by relevant stakeholders.

2.2.13 Behavior-Driven Development: A Development Practice

2.2.13.1 Information

Behavior-driven development (BDD) is a development method based on scenario descriptions. Note that BSI PAS has defined scenario as "a series of events involved that are linked to create a test or hazardous situation." One way to describe scenarios is to use the template used in Solis and Wang (2011). They use a scenario template described as follows:

Scenario n: <Scenario Title>
Given <Context> **And** <Some more contexts>....
When <Event>
Then <Outcome> **And** <Some more outcomes>....

When we need more than one context, event, or result, they can be connected by using one or more "And." Given the focus on events and results, the BDD template is often also used to specify tests. There are also tools available to use this template to automate tests, such as Cucumber and FitNesse.

Another way to construct scenarios is described in ISO PAS 21448:2019. The scenarios here are also intended for safety analysis—Safety of the Intended Functionality—SotIF.

2.2.13.2 Benefits

BDD combined with STPA or other safety analysis methods will improve safety analysis and verification also in agile development—see Wang and Wagner (2018).

2.2.13.3 Safety Adaptations

In BDD safety verification, to generate and test scenarios, we need the unsafe control actions (UCAs), the process variables, and the algorithms from the safety report. A simple example of a UCA from Wang and Wagner (2018): "During auto-parking, the autonomous vehicle does not stop immediately when there is an obstacle upfront."

BDD safety verification has two steps: (1) the business analyst and the safety analyst (QA) and the developer establish a meeting to generate test scenarios. In a BDD test scenario, we write the possible trigger event for the UCA in **When** <Event>. The other process variables and algorithms are arranged in **Given** <Context>. Then <Outcome> presents the expected behavior—a safe control action.

2.2.13.4 Agile Practices: A Summary

There exist more than 50 named agile practices. Several of these practices cannot be used as is when developing SCSW as they do not include mandatory process requirements needed for the development of safety-critical software. We have evaluated ten of the most relevant practices and described necessary add-ons and safety adaptions to ensure that important international standards like IEC 61508-3:210, ISO 26262:2018, and EN 501328:2011 are satisfied. All of these ten practices may contribute to shorten the time to market, to reduce costs, to improve quality, and to more frequent releases. Two extended agile practices have been suggested: the "backlog splitting" and "four questions."

2.3 Challenges and Agile Solutions

The three most important challenges for any software development project—agile or not—are requirement management, bad management, and bad communication. We have identified these three main challenges by performing literature surveys and surveys for both the aviation and railway domain—see Myklebust et al. (2017a, b) for references. The corresponding relevant practices used to help in solving each challenge are presented in the tables shown in Fig. 2.4 (Table 2.6).

Table 2.6 Practices needed to handle the four most important challenges

Challenges	Relevant practices
1. Requirement management • Ambiguities • Frequent changes • Insufficient • Addition of new requirements • Immature standards • Different interpretation of standards	1. Backlog 2. Backlog splitting 3. Safety story refinement—involve both the sprint team and alongside engineering 4. Backlog refinement meetings 5. Definition of done 6. Definition of ready 7. Hazard stories 8. Prioritize requirements/work list 9. Safety stories 10. Test-driven requirement 11. User stories 12. Security story 13. RDA (requirement-driven analysis)
2. Bad management • Lack of resources • Lack of executive support • Lack of planning • Lack of IT management	1. Backlog refinement meetings 2. Definition of done 3. Definition of ready 4. Frequent releases 5. Iteration planning 6. Prioritize requirements/work list 7. Release planning 8. Sprint planning 9. Time box 10. Test-driven requirement
3. Bad communication • Incomplete requirements • Lack of user involvement • Did not need it any longer • Unrealistic expectations	1. Backlog refinement meetings 2. Daily scrum 3. Definition of done 4. Definition of ready 5. Prioritize requirements/work list 6. Retrospective 7. Sprint review/demo 8. Team-based estimation 9. Test-driven requirement 10. The wall (similarities with scrum and Kanban boards)

2.4 An Alternative View

An alternative way to look at agile development is to focus on what to do instead of just looking at a set of practices. The following discuss six "what's" that will help build a well-functioning agile development team.

- *Collaborate with the customer*: The customer is satisfied when requirements are fulfilled, expectations are met, and wants and needs are gratified. An Agile team is in near-constant communication with the customer, clarifying expectations, collaborating on fixes, and communicating options not previously considered. This frequent interaction between the team and the customer is what promotes creativity and heightens quality.

- *Work together daily*: According to the Agile Alliance, a common pitfall among agile teams is to equate a group of people who work together with a "team." Teams, and teamwork, contribute to successful projects when they collaborate as a cohesive unit. Organizational science researchers identified six components of teamwork quality:

  - Communication
  - Coordination
  - Balance of team member contributions
  - Mutual support
  - Effort
  - Cohesion

  There is a direct relationship between teamwork quality, team performance, and project success. Valuing individuals and interactions means practicing teamwork daily. However, agile teams do not operate in a vacuum. They need to interact with business stakeholders on a regular basis to infuse the development process with business priorities and domain expertise.

- *Build projects around motivated individuals*: It takes motivation to push through an intense development cycle and get the work done right. Agile teams are passionate about their work, focused on the team goal, and supportive of each other. When there is trust and respect among peers, agile teams establish a rhythm to their fast-paced and predictable work. In Tuckman's theory of group development, teams pass through four stages—forming, storming, norming, and performing. As the team gears up, members who have been accustomed to working alone become more adaptable. They are willing to:

  - Take on the required roles
  - Form collaborative relationships
  - Adopt what the team deems to be the most efficient processes
  - Work reliably without management oversight

  "When they are performing at full capacity, agile teams constantly learn from one another; it is an empowering and motivating environment." Agile teams that deliver value at speed belong together. Nothing kills motivation faster than reassignment or redistribution of the members.

- *Convey information face-to-face*: Whether working through a knotty problem with a teammate or reporting on the day's accomplishments at a daily meeting, Agile team members prefer face-to-face communication. Information lost in a full email box or voice message queue slows or blocks progress. The daily meeting is one time the entire team connects to find out if there are any issues that could cause delays. This brief, face-to-face encounter demands team members be present and forthcoming. Admitting there's a hang-up and trusting teammates to rally to the cause do not come easily to some professionals. Face-to-face conversation opens channels and builds trust, making the Agile method sustainable.

- *Form self-organizing teams*: Self-organizing teams choose how they will execute the work and who will do what. They divide the work into increments that can be completed within each iteration and into tasks that can be completed each day. Management does not assign tasks or look over their shoulders. Unless members have extensive prior experience, Agile teams do not intuitively know how to self-organize and plan and execute an Agile software development project. It takes training, coaching, and mentoring to make an agile team. A team that is performing at full throttle still benefits from a mentor who can help them grow their skills.
- *Reflect on how teams can become more effective*: Agile teams routinely examine their performance and look for ways to do better. In fact, they are dedicated to continuous improvement. One scheduled time for reflection is the retrospective, typically held right after each development iteration. Team members share what went well and what went wrong and then identify how to improve the process on the next go-round. In a Scrum team retrospective, each member suggests something the team should start doing, stop doing, and continue doing.

2.5   Requirement Manager

Managing the requirements is an important activity and responsibility. Requirements come first—what shall we make?—and it comes last—have we done what we were supposed to do? It decides what the system shall do—functional requirements—and how the system shall do it—non-functional requirements. The table below shows the responsibilities for a requirements manager related to two important standards—EN 50128 (railroad) and IEC 61508 (generic) (Table 2.7).

2.6   Alongside Engineering

Alongside engineering is an activity invented during the SafeScrum project to handle all activities needed to complete a safety-critical development project but considered to be outside the realm of software development. Some of the activities done by the alongside engineering team are:

- Writing the safety plan.
- Writing the plans for verification and validation.
- Performing safety and risk analysis, both at the start of the project and each time there is a significant change to one or more requirements or to the system's operating environment.
- Writing and maintaining the agile safety case and the hazard log.

Table 2.7 Requirement responsibilities

EN 50128:2011/AC2020	Rewritten for generic safety including IEC 61508:2010
Responsibilities 1. Shall be responsible for specifying the software requirements 2. Shall own the software requirements specification 3. Shall establish and maintain traceability to and from system-level requirements like EN 50128:2011 4. Shall ensure the specifications and software requirements are under change and configuration management including state, version, and authorization status 5. Shall ensure consistency and completeness in the software requirements specification (with reference to user requirements and final environment of application) 6. Shall develop and maintain the software requirement documents *Key competencies* 1. Shall be competent in requirements engineering 2. Shall be experienced in application's domain 3. Shall be experienced in safety attributes of application's domain 4. Shall understand the overall role of the system and the environment of application 5. Shall understand analytical techniques and outcomes 6. Shall understand applicable regulations 7. Shall understand the requirements of EN 50128	*Responsibilities* 1. Shall be responsible for specifying the software requirements 2. Shall own the software requirements specification 3. Shall establish and maintain traceability to and from system-level requirements in relevant safety standards like IEC 61508 and, e.g., relevant maritime requirements, e.g., "safe return to port" 4. Shall ensure the specifications and software requirements are under change and configuration management including state, version, and authorization status 5. Shall ensure consistency and completeness in the software requirements specification (with reference to user requirements and final environment of application) 6. Shall develop and maintain the software requirement documents *Key competencies* 1. Shall be competent in requirements engineering 2. Shall be experienced in application's domain 3. Shall be experienced in safety attributes of application's domain 4. Shall understand the overall role of the system and the environment of application 5. Shall understand analytical techniques and outcomes 6. Shall understand applicable regulations 7. Shall understand the requirements of relevant safety standards like IEC 61508

- Performing safety validation at the end of each sprint. Thus, the RAMS—Reliability, Availability, Maintainability, and Safety—engineer is part of the alongside engineering team.

As we see, most of the activities done by the alongside engineering team are related to safety and security. In our experience, most developers are weak in these areas, mostly because they have learned little about it. The alongside engineering team may be considered a service team for the Scrum team—personnel who do the necessary tasks that the developers do not have time or necessary knowledge to perform. The diagram below illustrates the relationship between the alongside engineering team and the Scrum team (Sprint team) (Fig. 2.7).

To sum up—we need alongside engineering because:

- SW engineers want to develop code, not to read safety standards and write documents.

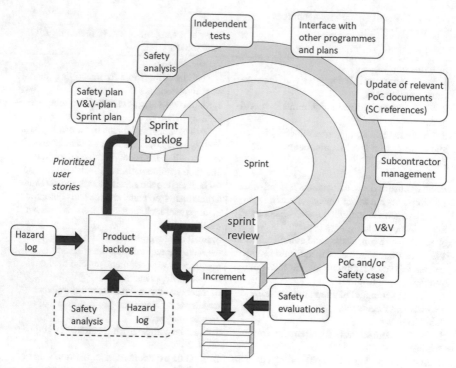

Fig. 2.7 How the sprint team and the alongside engineering team cooperates Extension based on Fig. 6.2 in Hanssen et al. (2018)

- There are many weak or missing topics in current safety and security standards such as weak process descriptions and weak deployment requirements. In addition, the standards are missing information on machine learning, artificial intelligence, and deployment over the air.
- Many issues related to "ordinary development" and development have to take into account imminent regulative aspects, e.g., autonomous levels, as part of regulation and safety standards.
- It can be used at the beginning of projects, before the Sprint team has been established.

By organizing a development project in this way, we allow both developers and safety experts to contribute to the project with what they are good at. However, this is not a recipe for creating knowledge silos in the organization. In order to understand the problems facing a development team, the persons belonging to the alongside engineering team need to be competent developers. In the same way, the developers need to have solid knowledge of safety—what it is, why it is important, and how to achieve it. In short—in order for the two groups to communicate efficiently, we need to build a safety culture among the developers.

The model for cooperation between the development team and the alongside engineering team shown above implies that the alongside engineering team will not

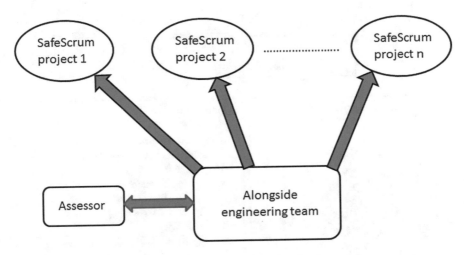

Fig. 2.8 The development projects and the alongside engineering team

work with the developers full time. Thus, under normal circumstances, the alongside engineering team will be able to service several development teams as shown in the diagram below.

In some cases, the development team also needs to communicate with the assessor. In our opinion, communication with the assessor would be most efficient if the alongside engineering team functioned as a liaison (Fig. 2.8).

An alongside engineering team is not only useful for an agile development project. For example, the frequent lack of safety-related expertise is not a problem only for agile projects. Thus, any software development project would benefit if the company had an alongside engineering team. However, this must not be an excuse for creating knowledge silos in the company—sharing a set of knowledge and "worldview" is important factors if we shall have efficient communication. An overview of the alongside engineering activities is shown in the diagram below (Fig. 2.9).

Instead of considering the necessary activities, an alternative way to describe the responsibility of the alongside engineering team is to look at roles involved and each role's responsibility. The following diagram shows the roles involved. We see that the roles belong to one out of the three areas of responsibility—the project itself, the alongside engineering, and the outside—which contains any third party and the assessor (Fig. 2.10).

The RAMS responsible is an important part of the alongside engineering team. He is responsible for the reliability, availability, maintainability, and safety of the system. Thus, we need someone with extensive knowledge of the relevant standards when it comes to safety and safety requirements. The RAMS responsible shall receive documents related to proof of compliance with the standard from both the Scrum team and from the rest of the alongside engineering team. He is responsible for verifying that all safety requirements are fulfilled or that there are reasonable reasons for any avoidance. Another important responsibility is to facilitate communication with the assessor. The RAMS responsible is thus the central resource on safety for the team, the Scrum master, the QA, and the product owner and should

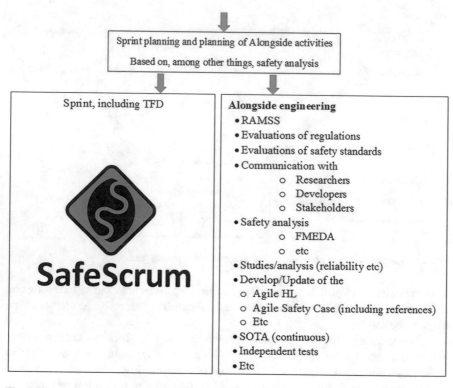

Fig. 2.9 An overview of alongside engineering activities

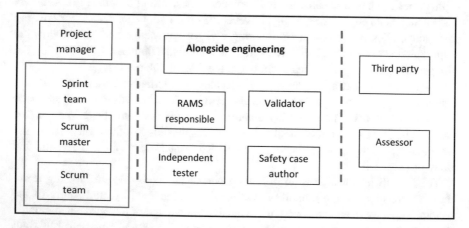

Fig. 2.10 Roles and responsibilities

take part in sprint planning, in sprint reviews, and in any type of discussions or clarifications that are needed to evaluate the meaning of safety requirements and how they are met by the solution that is being developed.

The role of independent tester and verifier is also an important part of the alongside engineering team. These testers need to be competent in the domain where testing is carried out and in existing test and verification methodologies. In addition, they must be able to identify the most suitable method in a given context. An independent tester must also be capable of deriving test cases from specifications, have the ability to think analytically, and have good observation skills.

The independent tester shall develop a software verification plan, stating what needs verification and what type of process (e.g., review, analysis, etc.) and test is required as evidence. He shall also check the adequacy—completeness, consistency, correctness, relevance, and traceability—of the documented evidence from previous reviews, integration, and testing. If there are anomalies, he needs to evaluate the risks related to these, record them, and communicate them to the change management body for evaluation and decision. The responsibilities of the independent tester also include managing the verification process (review, integration, and testing) and ensuring the independence of the required activities. In order to show proof of compliance, he must develop and maintain records on the verification activities and develop a verification report stating the outcome of the verification activities.

The validator must first and foremost develop a system understanding of the software within the intended environment of application and indicate any constraints as appropriate—what is called fitness for use. Based on this understanding, he shall develop a validation plan and specify the essential tasks and activities for software validation and get this plan accepted by the assessor. The software must be reviewed to check that it is meeting the software requirements. In addition, he needs to evaluate the conformity of the software process and the developed software against the requirements of the relevant European standards, including the assigned SIL. In order to achieve this, he needs to carry out audits, inspections, or reviews on the overall project—instantiations of the generic development process—as appropriate for all phases of development. It is necessary to review and analyze the validation reports relating to previous applications as appropriate. Last, but not least, he must check that the developed solutions are traceable to the software requirements.

All the test cases and verifications need to be reviewed for correctness, consistency, and adequacy. In addition, we need to ensure that all validation plan activities are carried out. Any deviation must be reviewed and classified for risk and other impacts. In order to include previous experiences, we need to ensure that the relevant hazard logs and remaining non-conformities are reviewed and that all hazards are closed in an appropriate manner through elimination or risks control/transfer measures. Finally, we need to write a validation report that shall state our agreement/ disagreement regarding the release of the software.

For large projects, we may consider two extra roles—safety case author and technical writer. None of these roles are defined in any standard or regulation. The safety case author is the person responsible for developing the safety case(s). In some projects, we may need different persons to develop different parts of the safety case. The role is sometimes combined with the role as RAMS manager. The technical writer prepares instruction manuals, articles, and other supporting documents to communicate complex and technical information more easily. He also develops, gathers, and disseminates technical information among customers, designers, and manufacturers.

References

AAMI TIR45: Guidance on the use of AGILE practices in the development of medical device software (2012)

Birk, A., Dingsøyr, T., Stålhane, T.: Postmortem: Never leave a project without it. IEEE Softw. **19** (3), 43–45 (2002)

Chaos: The Standish Group (1995)

Doss, O., Kelly, T.P.: Challenges and opportunities in agile development in safety critical systems: A survey. Softw. Eng. Notes. **41**(2), 32–33 (2016)

Eveleens, J.L., Verhoef, C.: The rise and fall of the Chaos report figures. IEEE Softw. **27**(1), 30–36 (2010)

FAA: Chapter 8: Safety analysis: Hazard analysis tasks. In: System Safety Handbook. FAA, Washington, DC (2000)

Hanssen, G. K., Stålhane, T., Myklebust, T.: SafeScrum—Agile Development of Safety-Critical Software. Springer, Cham (2018). isbn:9783319993348

Kniberg, H.: Scrum and XP from the Trenches, 2nd edn. C4Media, Toronto (2015)

Larman, C., Basili, V.R.: Iterative and incremental development: A brief history. IEEE Comput. **36** (6), 47–56 (2003)

Leite, I.M., Antonio, P.O., Nakagawa, E.Y.: From safety requirements to Just-enough safety-centered architectural solutions in agile contexts. SBES '20, Natal, 21–23 Oct 2020

Meyer, B.: Agile! The Good, the Hypa and the Ugly, 1st edn. Springer, Cham (2014)

Misra.: www.misra.org.uk/Activities/MISRAC/tabid/171/Default.aspx

Myklebust, T., Stålhane, T.: Safety stories—A new concept in Agile development. In: Fast Abstracts at International Conference on Computer Safety, Reliability, and Security (SAFECOMP 2016), Trondheim, Norway, Sep 2016

Myklebust, T., Stålhane, T.: The Agile Safety Case. Springer International Publishing, Cham (2018). isbn:9783319702643

Myklebust, T., Stålhane, T., Lyngby, N.: The agile safety plan. In: PSAM13, Seoul, 2016a

Myklebust, T. Stålhane, T., Hanssen, G.K.: Use of Agile Practices when developing Safety-Critical software. In ISSC 2016-08, Orlando, 2016b

Myklebust, T., Hanssen, G.K., Lyngby, N.: A survey of the software and safety case development practice in the railway signalling sector. In: ESREL Portoroz, Slovenia, 2017a

Myklebust, T., Bains, R., Hanssen, G.K., Stålhane, T.: The Agile Hazard Log approach. In: The 2nd International Conference on Engineering Sciences and Technologies, 2017b

Myklebust, T., Stålhane, T., Hanssen, G.K.: Analysis first development for agile development of safety-critical software. In: Conference: ESREL, 2019

Paasivaara, M., Durasiewicz, S., Lassenius, C.: Using Scrum in a globally distributed project: a case study. Softw. Process Improv. Pract. **13**, 527–544 (2008)

Ries, E.: The Lean Startup: How Today's Entrepreneurs Use Continuous Innovation to Create Radically Successful Businesses (2011)

Solis, C., Wang, X.: A study of the characteristics of behaviour driven development. In: 37th EUROMICRO Conference on Software Engineering and Advanced Applications, 2011

Spica.: www.spica.com/blog/timeboxing

Stålhane, T., Myklebust, T.: Hazard stories, HazId and safety stories in SafeScrum. In: XP 2018, Porto, 2018

Stålhane, T., Myklebust, T., Hanssen, G.K.: The application of Safe Scrum to IEC 61508 certifiable software. In: PSAM11/ESREL 2012. Helsinki, June 2012

Standish Chaos Press: Release on the Success of Technology Projects and Programs. Boston, MA, 2009

Stray, V., Sjøberg, D.I.K., Dybå, T.: The daily stand-up meeting: A grounded theory study. J. Syst. Softw. **114**, 101–124 (2016)

Vardhman, R.: 20+ Astonishing Agile Adoption Statistics for 2020. https://goremotely.net/blog/agile-adoption/ (21 Aug 2020)

VersionOne.: https://stateofagile.com/# (2020)

Wang, Y., Wagner, S.: Combining STPA and BDD for Safety Analysis and Verification in Agile Development: A Controlled Experiment. University of Stuttgart, Stuttgart (2018)

Chapter 3
POC in Agile Development and for SMEs

Learn from yesterday, live for today, hope for tomorrow. The important thing is not to stop questioning.

Albert Einstein

What This Chapter Is About
- Agile and Kanban
- Small SafeScrum teams
- What is an argument
- PoC for SMEs

3.1 Agile and Kanban

The main focus of Kanban is to accurately state the work to be done, when it is ready and when the developers have to perform the job. This is done by prioritizing tasks to be done and by defining the workflow and lead-time to delivery. Thus, a Kanban is strongly related to a prioritized project backlog. The most important item when using Kanban is the Kanban board. The board is a visual presentation of the project's status and shows:

- What is not done yet
- What is in process
- What is finished

Remember, there is no such thing as the "correct" Kanban board. It can contain a lot of information or almost no information at all as shown in Figs. 3.1 and 3.2. The Kanban board in Fig. 3.1 is adapted to the development of safety-critical software, and the next board in Fig. 3.2 is a simple Kanban board.

An important difference between Kanban and Scrum is that Scrum uses roles such as a product owner who sets product vision and priorities, the Scrum team who implements the product, and a Scrum master who removes impediments and

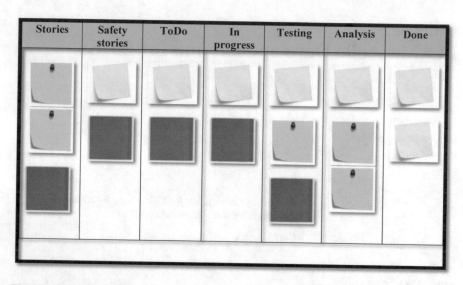

Fig. 3.1 Kanban board example for functional safety

Fig. 3.2 Kanban board

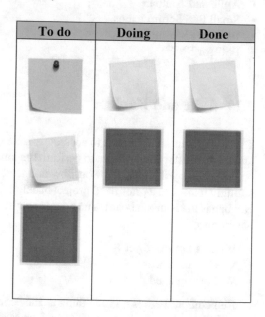

provides process leadership. Kanban does not prescribe any roles. Thus, you do not need to have a product owner role in Kanban. However, both Scrum and Kanban let you add whatever roles you need.

Scrum limits WIP (work in progress) indirectly, while Kanban limits WIP directly. The general mindset in both Scrum and Kanban is "less is more." So, when in doubt, start with less—i.e., just a few items in progress. Kanban says you should limit the WIP. There is only one way to identify this limit, and that is to experiment. Thus, start with a few items if that works well, then add a few items and run a new sprint. Keep on adding items to the sprint until you see that a new item might cause trouble. Note that this will only work if the sprint items are of roughly the same size.

An important difference between Scrum and Kanban is the rules pertaining to the Kanban Scrum board. In Scrum, the product owner cannot touch the Scrum board since the team has committed to a specific set of items in the iteration. A Scrum team may decide to allow a product owner to change priorities mid-sprint but this is considered an exception. In Kanban, we have to define rules for who is allowed to change what on the board. The product owner is given a "To Do," "Ready," "Backlog," or "Proposed" column to the far left, where he can make changes whenever he likes. A Kanban board does not need to be owned by one team since it is related to one workflow but not necessarily to one team. However, a team may decide to add restrictions about when priorities can be changed. The team may even decide to use time-boxed fixed-commitment iterations, just like Scrum.

Kanban and Scrum both have strong sides. Although interruptions are challenging for all projects, Kanban works "well" when handling interrupts and works fine also for large groups since communication and planning overhead are lower. Scrum excels at projects requiring deep collaboration and innovation and works best with small cross-functional teams.

Below are two sketches of how Kanban will allow us to handle the lack of resources in a software project—in this case, resources for testing.

- **Case 1**: Consider the following situation: project P1 needs more resources for testing. In order to achieve this, resources are moved from Analysis to Test. In order to achieve this, one analysis activity—A1—is set on wait. Having it explicitly marked as "waiting" is useful since we will not forget A1, and we can see what and how much is put on wait and see the state that A1 was left in (Fig. 3.3).
- **Case 2**: Another situation is that P2 needs more resources for testing. In this case, resources are moved from P1 Analysis to P2 Test. One analysis activity—P1-A1—is set on wait. This is useful since we will not forget A1, and we can see what and how much is put on wait. In addition, we see the state that A1 was left in (Fig. 3.4).

Some final words on Scrum and Kanban: By visualizing work in new ways, a Scrum team can apply the set of practices described here to more effectively optimize value delivery more efficiently.

There are, however, also other ways to handle interrupts. Berteig (2020) has some suggestions which are well worth mentioning. These are as follows:

1. Set aside some time for each Scrum team to handle interrupts. There may even be a separate Kanban board for this.
2. Negotiate. To quote Berteig (2020): "With this method, any time a stakeholder comes to the team with an interruption request, the Scrum Master/Coach/Process Facilitator writes the request on a bright collared note card so that it is easy to

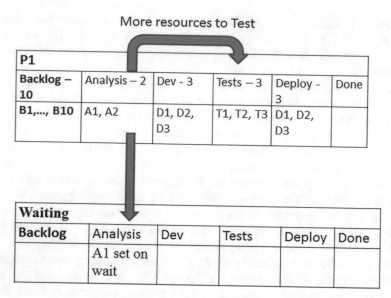

Fig. 3.3 Putting an activity on wait—Case 1

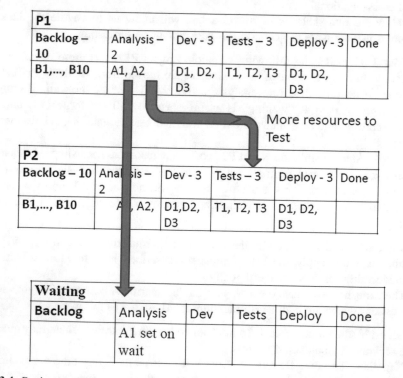

Fig. 3.4 Putting an activity on wait—Case 2

distinguish it from the other tasks the team is working on in their current cycle." The good thing with this approach is that the consequences both for the current and the new activities are brought into light and the consequences will be clear for all to see.

3. Have a separate team to handle interrupts. We will not recommend this since each project may require a unique combination of knowledge, and just having an "interrupt-team" available may not solve the problem.
4. Have short sprints. In this way, it will be just a short time before the team is available to handle the needed interrupt.

3.2 Small SafeScrum Teams: Less Than Seven Persons

3.2.1 Introduction

The purpose of this section is to look at problems and opportunities for start-ups when they want to start a SafeScrum project with a small staff—e.g., just three persons. Another relevant situation is when starting a larger project, but only two or three persons are yet allocated to the development team. The challenge for a SafeScrum team—large or small—is often the alongside engineering resources—RAMS responsible, independent tester, validator, and safety-case author. If the alongside engineering team is good, we have fewer problems. As a minimum, we need a RAMS responsible and a requirement manager. If these positions cannot be filled (these roles can be filled by one person), the company probably should not develop safety-critical software at all.

The Teams and Their Environments
Figure 3.5 shows the two environments that the Scrum team must relate to—the alongside engineering team and the outside, made up by any third party involved and the assessor.

There are a few additional challenges related to small Scrum teams. Most of them also apply to other small development teams. The most important challenges identified are that:

- The company at present has no Scrum master available for the project.
- The company has never had a validator as a separate role, only as part of the independent tester role
- The company has never used a requirements manager as a separate role (see also Myklebust and Stålhane 2018). This can be improved by having good general routines for requirement management and including agile practices like backlog refinement, establishing safety stories, etc.
- Scrum is weak on quality management. See Hanssen et al. (2016) on how the development team may solve this for agile and ordinary projects, i.e., include a QA (quality assurance) person in the sprint team. For smaller teams, a common

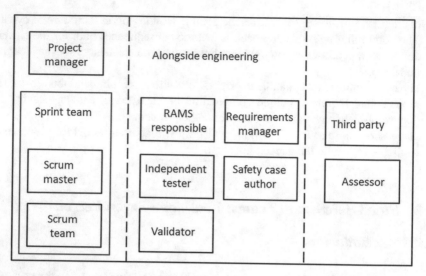

Fig. 3.5 Roles when developing safety-critical products and systems

QA person for both the sprint team and the alongside engineering is a pragmatic solution.

- A small Scrum team is critically dependent on the competence of all team members. There is no place to hide for a person with a lower competence.
- In complex projects, the team will not cover all the relevant technical topics. When developing safety-critical software, this can be critical for the certification of the product.

The Scrum Master

Being a Scrum master carries a lot of responsibilities. Some of the responsibilities are simpler than others. The following is just a short comment on each responsibility. A company planning to use Scrum or SafeScrum will need to see how each responsibility is handled in each project:

- Foster communication. Communication is the lifeblood of Scrum. However, you cannot force people to communicate—they must be convinced that they will profit from efficient communication if they are going to work together. Communication between the scrum team and the alongside engineering team is also of crucial importance.
- Protecting the team. In some ways, this is easy—just do not let management take people away from the Scrum team or give them extra work. However, saying No to management may be difficult and sometimes even risky. One way to protect the team is to make the rules clear to management before the project start. In addition, the cooperation between the SafeScrum team and the alongside engineering team should be clear.

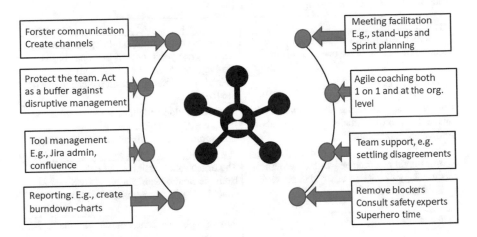

Fig. 3.6 A Scrum master's responsibilities

- Tool maintenance. This is not the most difficult part (after the tools have been validated)—make sure that the team has the tools they need and that they are updated when needed (as noncompliance is a very expensive solution).
- Reporting. This includes making burn-down charts, estimating the team's velocity, etc. This is not a complex job in ordinary projects. But reporting is a complex job by both the Scrum team and the alongside engineering teams when developing safety-critical software.
- Meeting facilitation. This is easy—make sure everybody participates with what they have done, what they will do, which problems they see, and if there are safety ramifications in any of the issues brought up.
- Agile coaching. This is an important point. To fulfill this role, the Scrum master needs to have a lot of knowledge and experience of Scrum. As a minimum, he must have worked in Scrum projects before—preferably in several projects.
- Team support. Some of this is easy while other things might require the ability to communicate with people efficiently.
- Remove blockers. This can be anything. Prepare for challenges. Some call it "Superhero time" (Fig. 3.6).

When browsing the internet, we find several descriptions and requirements for a Scrum master. Two issues are always present: soft skills—communications with other humans—and deep practical knowledge when using Scrum or SafeScrum.

Risks and challenges when applying Scrum.

A paper by Caballero et al. (2011) describes the risks clearly—SafeScrum parts are added. Note that quality in any form is the main concern (Table 3.1).

As a consequence of the problems identified in Berteig (2020), we need highly competent people in the Scrum team. We also need a validator and an independent tester since they are important roles in the alongside engineering team—the RAMS team. If the company does not have the necessary resources available, they need to

Table 3.1 Risks identified associated with Scrum

Risk	Action plan
The Scrum performance depends largely on the capability of involved team members	The team consists of, e.g., 4 engineers with 6 experience years. One of them has been working for 15 months with Scrum This also depends largely on the capability of the RAMS team
The weakness of Scrum is about quality management. It leaves too many things open about verification and testing	Code inspection and unit tests were inherited from, e.g., the previous process. The Scrum team should add a QA role (Hanssen et al. 2016)
Scrum should be combined with other agile methodology like XP in order to improve the verification practices	The organization assumes the risk Both the Scrum team and the RAMS teams should add relevant practices (Myklebust et al. 2018)
Self-managed team	Supported by the Scrum master and the RAMS manager

extend their staff or hire one or more consultants to fill the necessary roles. Alternatively, they may delay part of the work or hand it over to another team.

Both blogs and papers have identified a quality problem related to small Scrum teams. To quote Gfader (2020): "This means that the time for testing will be too short. Too little, done later, or tested in large batches (a lot at once), and that leads to blame, long test nights, late integration, bimonthly deployments, and general frustration." In addition, many companies also have too few automatic tests. This is especially a challenge for companies that develop safety-critical software.

For small teams, there have been experiments comparing Scrum and the Team Software Process (TSPi)—a pint-down academic version of the Team Software Process (TSP) that is a huge scale, cutting-edge, mechanical quality, a coordinated structure that aides improvement groups in delivering amazing programming concentrated frameworks. "TSPi method provides a series of operational processes to the software engineers, which can help them to achieve more efficient and effective software development and improve the quality and productivity of their software development project" (Jali et al. 2017).

Figure 3.7 shows that "PRO TSPI" apparently has better quality than "PRO Scrum" when applied to small teams. For teams using PRO TSP, only 3.8% of the defects have been detected in the operation phase, while for PRO Scrum, 9.3% of the defects are left for the operational phase. This ratio shows the quality profile of the process but does not determine the product quality level. The metric that let us know the product quality and effectiveness of each phase is the defect density. The developers consider a process effective when every phase has less or equal defect density than the last one.

The problems related to operational defects are best described in Gfader (2020) and can be illustrated by the following diagram and text:

"In my experience, that's what usually happens:

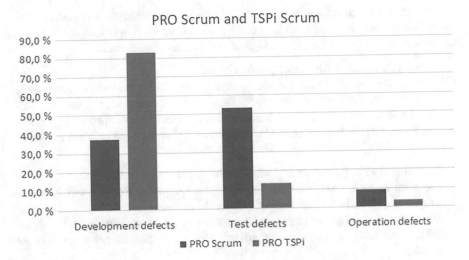

Fig. 3.7   PRO scrum and PRO TSPi

Fig. 3.8 The effect of too
few testing resources

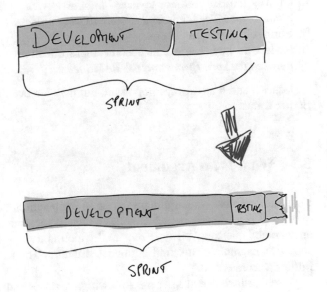

- Ahh... we're not quite done with development yet.
- Can you hold off on testing?
- Can't you just test in the next Sprint while we're still working?"

→ This should be the warning bell ringing by now.

That's usually what it looks like as a visualization:

This means that the time for testing will be too short. Too little, done later, or tested in large batches (a lot at once), and that leads to blame, long test nights, late integration, bimonthly deployments, and general frustration (Fig. 3.8).

Note that it is the RAMS manager together with the Scrum master and may be even the assessor who must keep the Scrum project from stumbling into this trap. If it is the assessor that does this, it is a sign of bad project management.

Important Practices with Small Scum Teams

Gfader (2020) has identified the six most important practices when doing agile software development with small teams. They are just common sense but still they seem to be forgotten or ignored. A more complete description of each agile practice can be found later in this chapter. Our comments following each practice are set in italics.

1. Collaborate with the customer—*who will be the customer in this case. This seems too often to be a challenge for agile in-house projects. For a small in-house project, e.g., the business manager will be the customer.*
2. Work together daily—*implies working places close to each other, either in an office landscape or neighboring offices. New tools like Teams may also be helpful.*
3. Build projects around motivated individuals.
4. Convey information face-to-face—*implies working places close to each other, either in an office landscape or neighboring offices.*
5. Form self-organizing teams.
6. Reflect on how teams can become more effective, *e.g., good communication between the sprint team and the RAMS team.*

Issues 5 and 6 are always important, but they are even more important for small Scrum teams.

3.3 What Is an Argument

First and foremost—an argument is not an explanation. An argument answers the question: "*How do you know*," which is a request for evidence, while an explanation answers the question: "*Why is that so*," which is a request for a cause. Table 3.2 shows an example of a sound argument structure. The terms used in the table are further elaborated.

A setting includes at least the following: A set of statements that are taken for granted by the manufacturer: the set of assumptions. A set of inference rules that are taken as acceptable for purposes of reasoning. Target is part of our notion of argument that it has a goal to establish some particular proposition. That is all that

Table 3.2 Argument structure

Refined argument		
Setting	Target	Reasoning structure
Rules and assumptions	Proposition to be established	Steps in the argument and relations among them

I mean by a "target" C, it is a proposition to be validated. A reasoning structure is similar to a derivation in logic; it consists of a sequence of statements that are meant to reach a target within a setting, with the statements annotated to identify them as premises, conclusions, or as inferred from others. Most of the detailed work of the theory of applied logic is to describe what a reasoning structure is and see how it bears on the success of the argument's success.

A successful argument is one in which:

- Every premise is among the statements assumed in the setting.
- Every inference is in accordance with a principle of inference assumed in the setting.
- The conclusion is the target identified in the goal.
- The reasoning structure is noncircular.
- There is no infinite regress of justifications for any step

Arguments in a PoC

In the definition of a safety case in the automotive standard ISO 26262-1:2018, it is defined as: *argument that functional safety is achieved for items, or elements, and satisfied by evidence compiled from work products of activities during development.*

Arguments are also an important part of PoC documents (references in the safety case). If the manufacturers already include the arguments in the PoC documents, it is less risky and much easier to finalize the safety case. In addition, it is easier for reviewers of the PoC document to evaluate the document. IEC 61508 does not require a safety case to be developed and presented, so in such cases it is especially important to include the arguments in the PoC documents.

The argument links safety requirements and objectives to the evidence and helps to explain this link. One should also be aware of argument defects and long argument texts that are presented to hide a bad design. The paper by Koopman et al. includes several relevant pitfalls related to argumentation, e.g.:

- Overlooking the safety relevance of sensors, actuators, software, etc.
- The "Checker" (e.g., the verifier) only covers a subset of the safety properties
- The common and obvious: violated assumptions
- Small changes
- Discounted failure(s)
- The human filter
- Insufficient testing and analysis, see FMEA example below

Below, we have listed the PoC documents that preferably should include arguments:

- Change impact analysis
- Hazard log
- Hazard, risk, and safety analysis
- Test plans and test reports
- Verification and validation plans and reports

- Proven in use analysis report
- Safety-related design principles

What is a valid argument?

The FAA System Safety Handbook (2000) includes information related to how to assess safety analysis, and the EASA Guide (2009) includes information related to the weaknesses of different techniques. Let us look at an example:

We have found all errors because we used analysis techniques and measures (T&M) named X.

Most manufacturers and assessors will probably accept this if method X is specified in the safety standard and only requires more information if another method is used. This is not the best approach and may not be sufficient in some cases.

FMEA example:

As an example, consider the T&M named "FMEA" applied to a printed circuit board (PCB). There is a considerable difference between just running an FMEA analysis—identifying failure modes without considering the limitations, deficiencies, and pitfalls—and just adding diagnosis (FMEDA). The first approach gives little confidence. The other that is described below includes all the considerations will give a lot.

Thus, "We have used FMEA" should not be considered a valid argument. What we need is something like "We have found all failures, identified the failure rates and calculated the failures for the undetected errors using relevant databases, etc.:"

- The discussions of the relevance of using this technique or measure for the product or system being analyzed are documented, and all relevant parts of the product or system have been analyzed—see document named X1.
- The participants have a long and documented experience using FMEA—see document named A.
- We have used the latest available, well-known sources for the failure rates—see reference list.
- We have used a validated calculation sheet when calculating the failure rates—see document named B.
- The FMEA process was performed according to IEC 60812:2018 (FMEA) and/or, e.g., IEC TR 62987:2015 (nuclear power plant—FMEA) with the add-on of evaluating the diagnostic functions.
- We used a template that was a basis for the previous certification—see document named C.
- The manufacturer has performed quality assurance of the FMEDA report.

In this case, we can inspect the FMEDA report and accept or reject the results. We can check the process, experience, and its results. Thus, this approach will build confidence, mainly because it is possible to check all the steps. Note that arguments need to be to the point, logical, and easy to understand. Long, complicated arguments will be difficult to read and understand. In addition, they may create an impression that someone is trying to hide something.

Evidence in a Safety Case

Evidence is not defined in ISO 26262. However, UL 4600:2020 has supplied a list of what they consider relevant evidence:

1. Experimental data.
2. Analytic data.
3. Procedure definitions.
4. Process compliance.
5. Development and V&V process data.
6. V&V data—e.g., test plans, test results, experimental design methodology, formal verification.
7. Qualitative analysis and subjective judgment.
8. Field engineering feedback data (especially important when having a DevOps approach).
9. Placeholder for evidence that will be collected via field engineering feedback.
10. Accepted risks, including evidence that risk is acceptable (this information is often included in the hazard log).
11. Assumptions for which no evidence is provided, including a basis of support that the assumption is reasonable.

Item 11 in the list above needs some clarification. The best way to do this is to give a small example: An assumption that traffic signals never display conflicting greens might lack field data. However, a discussion with a traffic signal engineer might be cited indicating that there are hardware fail-safes that disable the signal if conflicting greens are displayed, and the argument could be that this makes it reasonable to assume it will never happen in practice.

3.4 POC for SMEs

3.4.1 Introduction

When wanting to demonstrate Proof of Compliance, the first question is which standard you want to claim compliance with. In this section, we will assume that you claim compliance with ISO 9001:2015. In this case, some requirements must be met, irrespective of what type of company you are—small, medium, or large. These are the requirements related to the organization, top management—their responsibility related to leadership—and to quality policy.

An important issue in ISO 9001:2015 is "documented information," which is defined by the standard as "information that must be controlled and maintained. Therefore, it expects that you also maintain and control the medium as well as the information. ... A form is a document, when the form is filled out it becomes a record." The term "document" is not used in the standard anymore.

When creating and updating documented information, the organization must ensure appropriate identification and description such as title, date, author, or reference number and format—e.g., language, software version, graphics, and media—e.g., paper or electronic. In addition, it shall include information about review and approval for suitability and adequacy. Note that this is all that is required for documented information. Otherwise, it can be almost anything—a formal document, a memo, or a snapshot of a whiteboard. The manufacturers shall protect documented information retained as evidence of compliance from unintended alterations.

3.4.2   Quality and Safety Culture

ISO 9001:2015 has no requirements regarding a company's quality culture. Saftly (2012), in his article "Five essential ingredients for a quality culture," has listed up the following five points that make up a quality culture:

1. A mentality of "we're all in this together" (the company, suppliers, and customers). The company not just as the buildings, assets, and employees, but also customers and suppliers. The goal is consistently win-win-win for all parties.
2. Open, honest communication is vital. An important way to encourage truth-telling is by creating a culture where people listen to one another. This is a culture where open, honest communication is understood as necessary for people to function best.
3. Information is accessible. Information accessibility is at the heart of the work we do. Business leaders should be open about sharing information on the company's strategic goals because this information provides direction for what we will do next and—more importantly—direction for improvement.
4. Focused on processes. Everyone should move away from a "blame the person" mentality to a "blame the process and let's fix it" approach to problems and improvement.
5. There are no successes or failures, just learning experiences. An important insight is that failure and success are always value judgments we form after the fact. We can never predict with certainty whether what we do will end up as a success or a failure (or a mistake). We do the best we can based on our current experience, information, and understanding, and something happens.

Having a quality culture will not get you certified or make the assessor agree that you are ISO 9001 compliant. It will, however, help, and it will show the assessor that the most important things are in place and make him less obsessed with digging for details.

Both railways and the automotive industry have guidelines for a safety culture. In addition, the "data safety initiative working group" has published a checklist that the manufacturers (and assessors) can use to assess data safety—see DSIWG (Harney 2016). The European Agency for Railways has developed a European railway safety culture (see European Union Agency for Railways 2020). This model is made up of three building blocks:

- Cultural enablers: those levers through which an organizational culture develops.
- Behavior patterns: those shared ways of thinking and acting which convey the organizational culture.
- Railway safety fundamentals: those core principles which must be reflected by behavior patterns to achieve sustainable safety performance and organizational excellence.

ISO 26262-1 has defined safety culture as enduring values, attitudes, motivations, and knowledge of an organization in which safety is prioritized over competing goals in decisions and behavior. ISO 26262-2 has the following requirements for a safety culture: The organization shall:

- Create, foster, and sustain a safety culture that supports and encourages the effective achievement of functional safety
- Institute and maintain effective communication channels between functional safety, cyber security, and other disciplines that are related to the achievement of functional safety
- Perform the required safety activities, including the creation and management of the associated documentation in accordance with ISO 26262-8

3.4.3 Organization Requirements

The organization shall determine the processes needed for the quality management system and its application throughout the organization. According to ISO 9001:2015, section 4.4, the organization shall:

(a) Determine the inputs required and the outputs expected from the processes.
(b) Determine the sequence and interaction of the processes.
(c) Determine and apply the criteria and methods needed to ensure the effective operation and control of these processes. This includes monitoring, measurements, and related performance indicators.
(d) Determine the resources needed for the processes and ensure their availability.
(e) Assign the responsibilities and authorities for the processes.
(f) Address risks and opportunities.
(g) Evaluate the processes and implement any changes needed to ensure that they achieve their intended results.
(h) Improve the processes and the quality management system.

To the extent necessary, the organization shall also maintain documented information to support the operation of its processes and retain documented information necessary to give confidence that the processes are being carried out as planned.

3.4.4 Management Requirements

Even SMEs have top management—usually one person. Top management shall demonstrate leadership and commitment with respect to the quality management system. ISO 9001's requirement is that management shall demonstrate leadership and commitment with respect to the quality management system—see ISO 9001:2015, section 5.5.1—and

(a) Take accountability for the effectiveness of the quality management system
(b) Ensure that the quality policy and quality objectives are established for the quality management system and are compatible with the context and strategic direction of the organization
(c) Ensure the integration of the quality management system requirements into the organization's business processes
(d) Promote the use of the process approach and risk-based thinking
(e) Ensure that the resources needed for the quality management system are available
(f) Communicate the importance of effective quality management and of conforming to the quality management system requirements
(g) Ensure that the quality management system achieves its intended results
(h) Engage, direct and support persons to contribute to an effective management system
(i) Promoting improvement of the quality
(j) Support other relevant management roles to demonstrate their leadership as it applies to their areas of responsibility

Furthermore, according to ISO 9001:2015, section 5.2, top management shall establish, implement, and maintain a quality policy that is appropriate to the organization's purpose and context of the organization and supports its strategic direction and provides a framework for setting quality objectives. In addition, the quality policy must include a commitment to satisfy applicable requirements and to continual improvement of the quality management system. The quality policy shall be

(a) Available and be maintained as documented information
(b) Communicated, understood, and applied within the organization
(c) Available to relevant interested parties, as appropriate

Note that there is a lot of reuse opportunities when it comes to documented information. This holds both for complete documents and documents that are partly reusable. For the latter category, it is important that the documented information is structured with the goal of reuse in mind.

3.4.5 On Projects in SMEs

Projects in a small company usually have some distinct characteristics, first and foremost—they thrive on good communication. Everybody knows the other persons well and knows what they are doing and how they are doing it. SMEs usually have a common area with a coffee machine where a lot of communication takes place and people typically are on good terms. Factors that create effective communication, namely proximity—both emotionally, temporal and physical—are always available (see Sect. 1.4). Nothing beats face-to-face communication—preferably supported by a whiteboard or a large flip-over.

ISO 9001:20105, section 4.4, poses some challenges for SMEs, especially issue c—criteria for control and monitoring of the processes—and issue d—the resources needed. If you use the Scrum or SafeScrum process, however, this is simple. Control of what has been achieved in each sprint will give the control necessary.

Some organizations also seem to have problems with issue h—process improvement. The main problem here is data collection and analysis related to the efficiency of the process. However, if the company uses Scrum—or even better SafeScrum—they will get process improvement for free due to the requirement for sprint reviews.

An important effect of efficient communication is the need for few documents throughout the project. Problems are discovered, discussed, and solved without meetings and documentation. This is a problem when it comes to PoC—you may have done all the right things but there is no paper trail to prove it. A solution to the problem of the missing paper trail is to use a whiteboard when discussing something that needs to be documented by the development team. A snapshot of the whiteboard, a date plus a list of the participants will suffice as documented information according to ISO 9001. SMEs have challenges related to monitoring, measurement, performance indicators, and process improvement for the organization. The main reason for this is the need for data collection and analysis, which can be done by administrative personnel in a big company. In an SME, the developers must do this and is thus a direct drain on the manpower available.

A large part of the work related to PoC is about the process to agree on the architecture. A possible way to reduce this work is to (re)use architectural patterns. Using patterns is a way to reuse condensed experience, and the arguments for its usefulness should be based on "proven in use" arguments. For the proven-in-use argument to be valid, three conditions must be met: the old and the new environment, the use and the users must be similar. See also Sect. 8.5.1.

For PoC related to tests, the test log—input and output—will suffice. The only challenge is the relationship between requirements and test, which has to be documented. This documented information must show how each requirement has been validated—either via a test or via a step in the development process. See also Sect. 9.7.

References

Berteig, M.: https://berteig.com/how-to-apply-agile/seven-options-for-handling-interruptions-in-scrum-and-other-%E2%80%8Eagile-methods-3/, June 2020

Caballero, E., Calvo-Manzano, J.A., San Feliu, T.: Introducing scrum in a very small enterprise: a productivity and quality analysis. In: Conference Paper, June 2011

EASA Guide 2009. Safety Management System and Safety Culture Working Group (SMS WG) Guidance on Hazards Identification

European Union Agency for Railways: Introduction to the European Railway Safety Culture Model, 2020

FAA System Safety Handbook 2000. www.faa.gov/regulations_policies/handbooks_manuals/aviation/risk_management/ss_handbook/

Gfader, P.: 5 practices that help with agile software development, January 2020. www.scrum.org/resources/blog/5-practices-help-agile-software-development

Hanssen, G.K., Haugset, B., Stålhane, T., Myklebust, T., Kulbrandstad, I.: Quality assurance in scrum applied to safety critical software. In: International Conference on Agile Software Development (2016)

Harney, L.: Implementing the data safety guidance. In: 11th International Conference on System Safety and Cyber-Security (SSCS 2016), London, UK (2016)

Jali, N., Masli, A.B., Shiang, C.W., Bujang, Y.R., Mat, A.R., Hamdan, N.M.: The adoption of agile software methodology with team software process (TSPI) practices in the software engineering undergraduate course. J. IT Asia **7** (2017). www.invensislearning.com/blog/scrum-master-essential-skills-qualifications/

Myklebust, T., Stålhane, T.: The Agile Safety Case. ISBN 9783319702643. Springer International Publishing (February 2018)

Myklebust, T., Lyngby, N., Stålhane, T.: Agile practices when developing safety systems. In: PSAM14 Los Angeles, September 2018

Saftly, E.S.: Five essential ingredients for a quality culture (2012). www.processexcellencenetwork.com/lean-six-sigma-business-performance/articles/key-ingredients-for-quality-culture-development

Chapter 4
Generic Documents

It is not the strongest of the species that survive, nor the most intelligent, but the one most responsive to change.

Charles Darwin

What This Chapter Is About
- Documents and information management
- On living documents
- Change impact analysis
- Code baseline and configuration management
- Safety techniques and measures

4.1 Document and Information Management Plan

4.1.1 Definitions

First and foremost, we need to remember that documents and information are primarily about communication. Thus, you should read Sect. 1.4. We should start by defining the two terms: document and information. Due to its clarity and completeness, we have chosen to use the document definitions from Chen et al. (2005). "A document is a container of written information that the organization needs to track. It is created or received by an individual or an organization to undertake a business action. It is structured for multiple information users, and the authors have assembled the information in the document for human understanding. Manufacturers can store the documents and information in various formats such as paper and electronic, and media such as faxes, letters, computer hard drives or optical disks."

When it comes to the term information, things get complicated. To show you just *how* complicated, Chen et al. (2005) discuss six points of view, all summed up in the

© The Author(s), under exclusive license to Springer Nature Switzerland AG 2021
T. Myklebust, T. Stålhane, *Functional Safety and Proof of Compliance*,
https://doi.org/10.1007/978-3-030-86152-0_4

bullet list below. Even though we will not use any of the definitions in this list, it is important since it shows how many-faceted the term "information" is.

- Information as a resource
- Information as useful data
- Information as a representation of knowledge
- Information as part of the communication process
- Information as a constitutive force in society
- Information as understanding

Information as a constitutive force needs a little extra attention. It is important to understand that information carries the power of possible change. It may change peoples' opinions, preferences, or actions. This will, e.g., happen when we get information about new tools. The new tools may change the way we work and how, e.g., a project is managed. It may even change the whole company.

Chen et al. were not able to choose one of these views, and neither was this book's authors. Since we need a definition of information, we had to look elsewhere. Porat (1977) has given a short and precise definition which we will use:

> Information is data that have been organized and communicated. The information activity includes all the resources consumed in producing, processing and distributing information goods and services.

IEC 61508-1:2010 has taken the easy way out by bundling information and documentation. They state in a note to clause 5.1.2 that "The documentation requirements in this standard are concerned, essentially, with information rather than physical documents. The information need not be contained in physical documents unless this is explicitly declared in the relevant subclause." This approach is also taken by ISO 9001 when they state that "organizations can choose to use terms which suit their operations, e.g., using "records," "documentation," or "protocols" rather than "documented information"; or "supplier," "partner," or "vendor" rather than "external provider." To quote a slightly modified Hamlet, "There are more things in heaven and earth, Horatio, than are dreamt of in your standards."

4.1.2   The Plan

To quote ISO/IEC/IEEE 15288:2015: *"Information management plans, executes, and controls the provision of information to designated stakeholders that is unambiguous, complete, verifiable, consistent, modifiable, traceable, and presentable."*

Note that the information and documentation management plans are part of the project plan, and as such, they need to be assessed and controlled. See also the section on the agile safety plan—Sect. 5.1.

Qatar National Project Management (2021) has presented a list of issues that should be covered by a document and information plan. The list is short and to the point, and we will thus use this as our starting point in this section. However, not all items on the list are considered to be relevant for this section. We will focus on the following issues: naming conventions, storage, recovery and backup, security, and approval.

Naming Conventions

Naming conventions are important to quick identification of a needed document. A useful document name needs to at least contain the following information:

- Main contents—terms like "System-A" or "X-Contract"
- Type of document—terms like "test," "system-doc," etc.
- Version of the document—e.g., v.1
- Status of the document—e.g., "final" or "for-review"

Thus, a document could be named as follows "System-A_test_v.3_for-review." The interpretation of this document title should be self-explanatory.

Storage

All documents and information will have to be stored somewhere. Will this be on each person's PC, on a project server, or on a company server? In addition, it is quite common for people to keep a version on their PC that they are working on, which most likely will be different from the one they have stored on the project server.

The company needs to have rules and regulations for this—e.g., the project or company server shall always have the latest version at the end of the day. The decision needs to be coupled with the recovery and backup rules.

Recovery and Backup

All document and information plans need to have a backup strategy. As a minimum, this should include

- Who will create back-up file and when—e.g., every time something is changes or once a week
- How will the back-up files be labelled
- Where will the back-up files be stored—on site, somewhere else, in a fireproof safe or just on shelf
- Who will have access to the backup files
- What is the routines for document recovery—e.g., who can ask for it

Security

Security is a double-edged sword. We want the documents to be secure so that no unauthorized person can get access. On the other hand, we want a system where it is easy to give access to everybody who needs it. It is necessary to provide a system of multilevel security, e.g. full read and write access or read-only access. In addition, it should be possible to have different security levels for different parts of the document. This is especially important when we use external consultants or researchers who need access to some of our documents.

In addition, we need security rules for not-finished documents, which are often kept on the PC by each employee.

Approval

The company also needs to state which documents that need approval, and which do not. Company-related finished documents will need some kind of approval. The document and information management plan should specify which position in the company is allowed to approve which type of documents. Not-finished documents will need no approval.

4.2 On Living Documents

This section starts by discussing live documents in general. We then go on to discuss two important living documents: the change log and the code base.

One of the most important living documents is the change log. The change log contains information on all changes done to a system—when, by whom, and why, plus a reference to the proof of compliance for the changes performed. For changes due to DevOps feedback, it is also important to register the operator who reported the error or need for new functionality and the system and operational environment that were used.

We have several definitions of a living document. Wikipedia defines a living document as "A living document, also known as an evergreen document or dynamic document, is a document that is continually edited and updated." However, this definition cannot be used for developing, maintaining, or operating a safety-critical system. We will instead use a definition from Course Hero (2021), which states that a living document is "A type of document that is created as a draft document and has a *defined mechanism and process to continually review, monitor, revise, edit and update the document with defined and communicated accurate and just authority, processes and rules for the person who has the role, responsibility and authorization to manage, review, monitor, revise, edit and update the specific document* and date each version of the document and keep a log of each form of the document throughout the process and the list of people who subscribe to or need confirmation of any changes or revision of the document."

For a living document related to a safety-critical system, the important part of the definition above is in bold and italics. This implies that in order to have a living document related to a safety-critical system, you need a defined process and rules regarding management, reviewing, revising, and editing the document.

One of the ideas behind living documents is that they always are updated with the latest available information. Introducing proof of compliance will slow down the idea of living documents.

In order to provide proof of compliance, you need all the related processes to leave a paper trail—e.g., who edited the document, how it was done, and what the editing was based on.

The PoC for any living document must contain references to the document, the changes—date, reason, and who was responsible for the changes.

4.3   Change Impact Analysis Report

4.3.1   Introduction

Change impact analysis of the safety of products or systems is used by companies in many industries and is required by several standards. The CIA report (CIAR) is one of the main inputs to the assessor. A standardized CIAR will simplify the work both for the manufacturer, operator, and assessor and will also improve the certification process.

The guidance for a CIA plan and report is intended to be complementary to the standards and plans. The guidance for a CIAR will ensure that the manufacturers or operators document the CIA in such a way that an assessor will accept the report. It will also ensure that the modifications performed are sufficiently thought through by the manufacturer. Several examples exist where this has not been the case, and an inadequate CIAR had resulted in products that either have not been approved or the time before the product was on the market has been delayed with months and even years.

4.3.2   Input Documents and Related Plans

Below are the relevant input documents listed together with related plans (Table 4.1).

4.3.3   Minor Safety Issues and Relevant Process

The RAMS engineer is closely involved in the process to uncover and resolve minor safety concerns as early as possible. However, he or she is only allowed to resolve minor issues (e.g., issues not resulting in a change of the SRS) to the software being developed. Within a sprint, there are two points in time where the RAMS engineer may assist the sprint team in assessing safety. Firstly, in the sprint planning meeting, stories are selected and added to the sprint backlog. Since there may be one or more

Table 4.1   Input documents and related plans

Input documents	Related plans
• Relevant safety standards such as e.g., IEC 61508, ISO 26262, and EN 5012x series • Description of the system (DoS) • Description of the ODD including OEDR • FRACAS (failure reporting, analysis, and corrective action system) • Bug reports and similar • Change requests	• Project plan • Safety plan, see Sect. 5.1 of this book • Safety plan issued by suppliers • Release plan, see Sect. 8.3.1 in this book • IID (iterative and incremental development) plan, see Sect. 2.1 in this book

common actions (e.g., safety story) to be implemented or updated, we need to do the part of a detailed design that will affect more than one user story. The RAMS engineer should be present to review suggested design ideas and to assist the team.

Secondly, the RAMS engineer should participate in the sprint review meeting when resolved stories are demonstrated and reviewed. His or her role is to check that what was implemented by the software engineers in the last sprint is OK regarding system safety and safety requirements implemented in the last sprint. This will be a valuable support to the product owner responsible for approving stories as done (Fig. 4.1).

Topics and Chapters of the CIAR

Topic 1: Distribution list

This topic should be in line with ISO 9001:2015 "4.2.3 Control of documents...f) to ensure that documents of external origin determined by the organization to be necessary for the planning and operation of the quality management system are identified and their distribution controlled."

This also applies for safety management. As an example, controlled distribution of safety management documents could be distributed to the RAMS/safety manager, compliance manager, validator, assessor, and other relevant stakeholders. In some assessment projects, the relevant documents submitted to the assessor are presented separately or in an Annex of a document list.

Topic 2: Names of authors, reviewer(s), and signatories

No further explanation needed.

Topic 3: Revision history, including the current version number

Summarize the change in a few sentences. Version number and date have to be included. This is both practical and important information for assessors and employees. Relevant changes since the last edition should be described in the CIAR.

Topic 4: Status

Example "Draft," "Consultation," "Released," or "Expired."

Topic 5: Table of content

No further explanation needed

Chapter 1

Purpose,

   Scope,

   Context information,

   Project information: e.g., these changes are related to the current release plan.

   Relevant standards: e.g., test, safety, and security standards.

   Definitions: Sometimes only reference to a Definition document or database.

Chapter 2

Name of CIA participants, including information related to competence and experience.

Selection of relevant and sufficient number of experts is an important part of an Impact analysis.

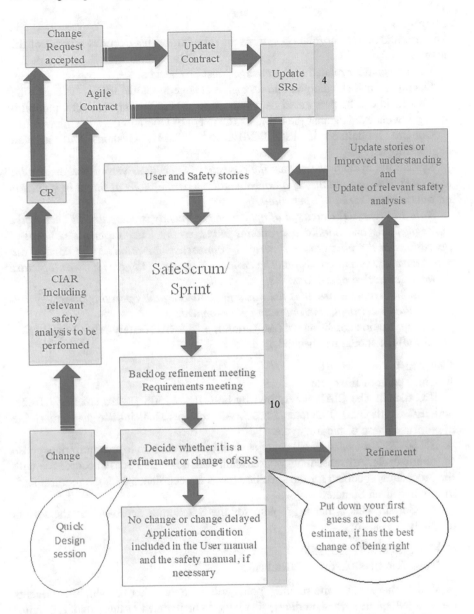

Fig. 4.1 Shows a relevant process when dealing with minor modifications. Number in grey indicates the IEC 61508:2010 safety phases. Extension of this figure based on Fig. 8.2 in Hanssen et al. (2018)

The information of the experts involved is often included as part of other chapters in the CIA, e.g., the chapter containing the names of the participants, analysis dates, meeting days, etc.

Chapter 3

Any deviations from normal operations and conditions that occur as a result of this change.

Failure behavior related to the change must be checked.

Operations can be changed due to, e.g., a change in ODD/OEDR.

This topic can be performed as part of, e.g., a HazOp (Hazard and Operability studies) or checked against the ConOps (Concept of Operation).

ConOps is defined in ISO/IEC/IEEE 15288:2015 Systems and software engineering as:

System life cycle processes as: *the concept of operations verbal and/or graphic statement, in broad outline, of an organization's assumptions or intent in regard to an operation or series of operations.*

Note 1 to entry: The concept of operations frequently is embodied in long-range strategic plans and annual operational plans. In the latter case, the concept of operations in the plan covers a series of connected operations to be carried out simultaneously or in succession. The concept is designed to give an overall picture of the organization operations.

Note 2 to entry: It provides the basis for bounding the operating space, system capabilities, interfaces, and operating environment.

The application condition list (AC) and/or, e.g., SRAC (safety-related application condition) list should be checked.

Chapter 4

Re-entry point of life cycle

IEC 61508: The CIAR is required by both IEC 61508-1:2010 chapter 7.16.2.6 and IEC 61508-3:2010 chapter 7.8.2.5. All changes shall initiate a return to the appropriate phase of the life cycle.

EN 50128:2011 6.6.4.2: All changes shall initiate a return to an appropriate phase of the life cycle. All subsequent phases shall then be carried out in accordance with the procedures specified for the specific phases in accordance with the requirements in this European Standard.

SafeScrum: This is not a problem for the SafeScrum approach, as the sprints normally are part of

- Phase 10 in IEC 61508:2010
- Phase 6 in EN 50126:2017 (Railway)

Which phase that is the re-entry point will have an effect on which documents (or e.g., automotive work products) that have to be updated or upgraded. Examples are given in EN 50128:2011 chapter 5.3 "Lifecycles issues and documentation."

Chapter 5

Required analysis, tests, verification, and validation

Describe the analysis, verifications, and validation steps required. This can normally be based on reviewing the verification and validation plan, including evaluation of required updates to T&M (Techniques and Measures). Current safety standards are weak when it comes to safety analysis of software. IEC will improve this in the next edition of IEC 61508. Even SW may have an influence on EMC performance (Chen et al. 2005).

Perform regression-based selective retesting of a system or component to verify that those modifications did not have unintended effects and that the system or component still complies with its specified requirements. Also, consider additional tests as it is not always sufficient to do the originally planned regression tests. Currently, safety standards are weak when it comes to regression testing requirements. IEC will improve this in the next edition of IEC 61508.

When applying SafeScrum and the sprints are completed by the sprint team, a final RAMS validation needs to be done by, e.g., the RAMS team. Since most of the system has been incrementally validated during the sprints, we expect the final RAMS validation to be less extensive than when using other development paradigms. This will help us to reduce the time and cost needed for assessment and certification. An extra decrease in assessment cost is expected when test-first-driven development is used.

See also IEEE 1012:2012, System, Software, and Hardware Verification and Validation and ISO/IEC/IEEE 29119-3:2013 Software and systems engineering—Software testing—Part 3: Test documentation.

EN 50128:2011 requires "5.1.2.8 *The Validator shall give agreement/disagreement for the software release.*"

Chapter 6
Assessor, certification, and authorization aspects

When the modified product or system has to be recertified by the certification body, one may consider using a new certification body or assessor for the modification. For example, new competence may be required by an assessor with special knowledge of a domain or country/region.

Established suppliers to the safety product market are most likely familiar with systems with complex requirements included in the safety standards. However, industrial suppliers and start-up companies new to the safety market may underestimate the extent of component and system qualification efforts needed. Sometimes companies, especially small companies, wish to expand their geographical scope. They have to check whether selling to more countries may affect the authorization. Other aspects, such as how the evidence shall be presented, may also be affected. Suppose the new country is, e.g., the UK. In that case, it could be wise to develop a safety case for most of the domains including domains that normally do not require a safety case—e.g., automotive trial operation.

Suppose several countries are included in the sales plan. In that case, it may be wise to apply for an IECEE certificate (www.iecee.org/certification/certificates/) for some standards, especially related to normative standards referred to by the safety standards.

The manufacturer should discuss challenging interpretations of the standard requirements related to the upgrade or new design with the assessor at the beginning of the project. The IEC System for Conformity Assessment Schemes for Electrotechnical Equipment and Components (IECEE) Certification Body

(CB) Scheme is the world's first international system for mutual acceptance of product safety test reports and certificates for electrical and electronic equipment, devices, and components.

Chapter 7

Required document/information changes

All affected documents shall be updated—see the "Living document" chapter. The documents that have to be updated should be identified in the CIAR. The relevant documents that have to be updated are usually listed in the "Document plan," "Safety Case," and/or the "CER (Conformity Evidence Report)."

The CER Method: This method is based on the IEC TRF (Test Report Format) method, as described in Worldwide System for Conformity Testing and Certification of Electrotechnical Equipment and Components (IECEE Guides 2021). The IEC TRF system is intended to facilitate certification or approval according to IEC standards. The TRF and CER method seeks to help the industry avoid unnecessary obstacles to trade and encourage different countries to harmonize their national standards and certification activities.

Chapter 8

Conclusion/summary

The conclusion of a CIAR summarizes the content and purpose of the analysis. The conclusions should be precise and to the point. The conclusion should briefly state the implications of the analyses.

The CIAR should

- Answer the question: Why should the assessor believe your result?
- Show evidence that your result is valid or why it will be valid—e.g., that the CIAR helps to solve the problem related to the process or system you should solve according to the change requests or similar.

Chapter 9

Document references or, e.g., ISO 26262 work products

Relevant documents/information to be updated and, e.g., new documents to be developed.

See, e.g., Annex A of this book.

4.4   Code Baseline and Configuration Management

The IEEE 24765:2010 standard defines the allocated baseline in configuration management as "the initial approved specifications governing the development of configuration items that are part of a higher-level configuration item. Related terms not handled here are developmental configuration, functional baseline, product baseline, allocated configuration identification."

NASA (2013) defines software configuration management as the process whose objective is to identify the configuration of software at discrete points in time and the systematic control of changes to the identified configuration to maintain software integrity and traceability throughout the software life cycle. In order to accomplish this objective, there are four identified SCM (Software Configuration Management) functions:

- Identification of the components that make up the software system and that define its functional characteristics
- Control of changes to those components
- Reporting of the status of the processing of change requests and, for approved requests, their implementation status
- Authentication that the controlled items meet their requirements and are ready for delivery

ISO 26262-1:2018 defines a baseline as a "version of the approved set of one or more work products, items" or elements that serve as a basis for change. According to the same standard, an element can be a system, a component, a hardware part, or a software item. Note that ISO 26262 defines an item as system or combination of systems, to which ISO 26262 is applied and implements a function or part of a function at the vehicle level.

IEC 61508-4:2010 defines a baseline as "Configuration baseline information that allows the software release to be recreated in an auditable and systematic way, including: all source code, data, run time files, documentation, configuration files, and installation scripts that comprise a software release; information about compilers, operating systems, and development tools used to create the software release."

EN 50128:2011/AC2020—a railway standard—defines a software baseline as a "complete and consistent set of source code, executable files, configuration files, installation scripts and documentation that are needed for a software release. Information about compilers, operating systems, pre-existing software, and dependent tools is stored as part of the baseline."

Keeping the source code under control through configuration management (CM) is a "must" for software certification, development, and maintenance. If all your customers only ran the same latest version of your system, configuration management would be needed only in one case—when we need to roll back the software due to an incorrect or incomplete change. This will be more important as DevOps become more popular. When applying the DevOps process, we need to be able to recreate the version of the system that the customer used when reporting the problem. This is necessary in order to be able to reproduce the reported problem.

When using or referring to the IEC 61508 standard, note that this standard uses the term "item" synonymously with "functional unit."

There are at least two applicable standards: IEEE 828:2012, standard for systems configuration management, and ISO 10007:2017 which contains guidelines—not requirements—for the system configuration part of QA. In addition, several safety standards also handle software configuration control—e.g., ISO 26262-6:2018, annex C (informative), which is dedicated to SCM and IEC 61508-3 with its section on additional requirements for SC management of safety-related software.

The railway standard EN 50128:2011 has two parts that should be considered in this chapter annex B10 (normative), which is normative and specifies the configuration manager's role, and annex C (informative), the document control summary. The components of a software system that are controlled by the SCM process include project and product documentation, all software code, and any document needed to meet a set of requirements or contractual obligations. In short, anything that is related to what we are contractually obliged to deliver to a customer. The new configuration will contain changes from the previous configuration. Thus, we need to see the test reports to ensure ourselves that both the new or corrected features and the old features work as intended.

Release planning is important for configuration management. Unfortunately, neither ISO 26262 nor IEC 61508 mentions the term "release planning." ISO 26262 shortly mentions components released for production, and EN 50128 discusses issues related to release in section 9—Software deployment and maintenance. ISO/PAS 21448 discusses methods and criteria for a software release, with a focus on the safety of the intended functionality (SOTIF) for road vehicles. We believe that in the future release, also for safety-critical systems, will be more frequent, due to both DevOps and more use of agile methods.

SWEBOK (2015) gives a complete description of the process needed for SCM. As in all other cases of PoC, we must provide evidence that everything is done according to the rules. According to the SWEBOK, a complete SCM must contain the following:

- Management of the SCM process—first and foremost, you must have a documented and implemented SCM process.
- SC identity—a list of the items that are to be controlled
- SC control—change control, change impact analysis
- SC status accounting—report what has been done and what is the status
- SW release management and delivery—building and releasing a new version

SC control and status accounting are important since it, among other things, is concerned with three important questions that need to be answered:

- When should we make a new version? There are several strategies used in the industry—e.g., at fixed dates, when a defined number of updates have been made or when an especially critical error has been corrected.
- Given that we have decided to release a new version, who decides which updates go into this version? Several persons could make this decision—e.g., the project leader, the customer relation manager, or the company manager. The important point is that it has to be decided who it will be.
- What should be included in the new version, and what should wait until a future release? If we can include all available changes/error corrections, there is nothing to decide here. However, new problems will be reported, new features will be needed, or there is already a backlog of problems waiting to be dealt with. The

whole process is the problem of assigning priorities—which change will be included when?
- Since different customers have different wishes and needs, we will most likely have to maintain several versions, making SCM one of the most important processes in the company.
- Without a real SCM process, a company will sooner or later deliver the wrong updates to one or more customers with potentially catastrophic consequences.

Note the difference between release and version. First and foremost: a build is an executable file that is handed over to the tester to test the functionality of the developed part of the project. Then, a release is what is handed over to the client of the project after the development and testing phases are completed while a version is the number assigned to the release made according to the addition of the requirement of the client.

IEEE 828 has a slightly different approach when it comes to managing the SCM process. Their main points are as follows:

- Manage implementation of CMP (Configuration Management Plan). The company shall provide relevant training, necessary resources, and the needed tools and working environments.
- Monitor CM activities. We need to monitor resources used and the progress in the work vs. schedule. This includes scheduled deliveries. In addition, we need to monitor any possible new risks. The influence of necessary changes must be assessed.
- The CMP must be updated when information that affects the CMP changes. As a minimum, the plan should be reviewed periodically. Proposed changes must be evaluated, and all CMP changes must be communicated to the project team.

Note especially the requirement to consider the risks related to changes in the SCM process or planning.

We need the following reports in order to control and document the configuration management purposes. A report should contain:

(a) A list of configuration information included in a specific configuration baseline
(b) A list of configuration items and their configuration baselines
(c) Details of the current revision status and change history
(d) Status reports on changes and concessions
(e) Details of the status of a delivered and maintained configuration (e.g., part and traceability)

The firm Software Engineering Process Technology (SEPT) has produced a set of useful templates for SCM documents. The documented can be downloaded for free from the SEPT homepage (http://www.sept.org/).

4.5 Safety Techniques and Measures: Software

4.5.1 Introduction

The safety standards that are part of this book include several methods, techniques, and measures to ensure that risks:

- Are avoided or are less likely to occur
- Mitigated or controlled
- If they occur, their consequences have less impact

In ISO 26262:2018, Techniques & Measures (T&M) are named methods, while in IEC 61508:2010 and EN 50128:2011, they are named Techniques & Measures. Only ISO 26262: 2018 includes controllability. Controllability is the ability to avoid specified harm or damage through the timely reactions of the persons involved, possibly with support from external measures. The safety standards recommend T&M appropriate for each defined SIL. However, selecting T&M from the list of recommendations does not guarantee that the required safety integrity will be achieved.

The T&M specified in the annexes in IEC 61508-3:2010, and EN 50128:2011 are sound software engineering principles. Problems when developing safety-critical software are not caused by a lack of adherence to the standard per se but by ignorance of sound engineering principles related to the specified techniques. Note, some critical T&M is not included in the current standards. See the reference list.

In our opinion, the essential things needed when making safety-critical software are general, sound engineering competence, combined with competence in software development and the application domain, good communication within the development team, and mindfulness of safety. We believe that standards such as IEC 61508-3:2010 are helpful since they list more or less the best T&M available when the standards were issued.

Different project development contexts may give rise to different T&Ms. In such cases, it is allowed to choose other relevant T&Ms as long as a rationale exists, and the assessor accepts the rationale. For new methods not listed in the standard, we recommend consulting the safety assessor.

4.5.2 Requirements and Relevant Process

The safety standards require that appropriate T&Ms shall be used. Some T&Ms are highly recommended for the appropriate safety integrity level. With each T&M in the tables, there is a recommendation for safety integrity levels 1–4. If a T&M is not used, then the rationale for this should be detailed, considering that a large number of factors affect software systematic capability. The rationale should include information that ensures that the life cycle phase(s) requirements and objectives have been met. It is impossible to give an algorithm for combining the T&M that will be correct for any given application. Deciding which T&M that shall be applied as part of the development process should be performed during the safety planning and agreed upon with the assessor.

Table 4.2 lists the symbols used together to explain and how they are used in the different standards.

Figure 4.2 shows how the process can be when deciding which T&M to be used. The developers should choose techniques which are appropriate to their:

- Working procedures
- Company internal standards or change the internal standards
- Level of technical expertise

An alternative is to improve or include new or external expertise.
Other relevant T&M sources are:

- NASA software engineering benchmarking study (NASA 2013)
- NASA software safety guidebook (NASA 2004)
- MOD (Ministry of Defence) MIL-STD-882F (draft 2021)
- STAMP (Systems-Theoretic Accident Model and Processes) (Leveson 2004)
- STPA (Systems Theoretic Process Analysis) (Leveson 2015)
- Data-related references: SCSC guide (2021) and Myklebust et al. (2021)
- LOPA (Layer of Protection Analysis) IEC 61511-3:2016

Table 4.2 Symbols and relevant explanation

Symbol	Explanation	IEC 61508:2010	ISO 26262:2018	EN 50128:2011
M	This symbol means that the use of a technique is mandatory.	NA	NA	Used
HR	This symbol means that the technique or measure is Highly Recommended for this safety integrity level.	Used	Similar to "++"	Used
PR	Passive recommendation replaces "-" in edition 3 of IEC 61508	Replaces "-"	Similar to "o"	Use "-"
R	This symbol means that the technique or measure is Recommended for this safety integrity level.	Used	Similar to "+"	Used
NR	This symbol means that the technique or measure is positively Not Recommended for this safety integrity level.	Used	Similar to "o"	Used
"-"	This symbol means that the technique or measure has no recommendation for or against being used,	Similar to "---"	NA	Used

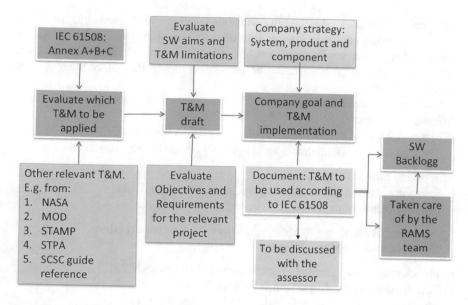

Fig. 4.2 Process for deciding which T&M to use

4.5.3 T&M and the Safety Life Cycle

Safety standards define a safety life cycle which is often a V-model. Figure 4.3 presents the IEC 61508 safety life cycle and relevant hazard, risk, and safety analysis.

There is no direct link between the safety life cycle and T&M for all phases.

Below, we have presented links between the safety lifecycle phases 1–10 and related T&M.

1. Concept Phase

In the safety standards, no tables listing T&Ms are listing T&M directly related to concept in IEC 61508-3:2010, ISO 26262-3:2018, and EN 50128:2011. During the last years, it has become more relevant for manufacturers to develop products and systems based on new concepts due to the rapid technology development. As a result, more emphasis has been put on this phase. Therefore, which T&M to be used in this phase that is relevant has become more critical. We have also seen that it may take less time from concept to finalized products.

Relevant T&M:

- Concept FMEA (Myklebust et al. 2018)
- HazId for different concepts (including ConOps) and different use cases
- STPA (Young and Leveson 2014)

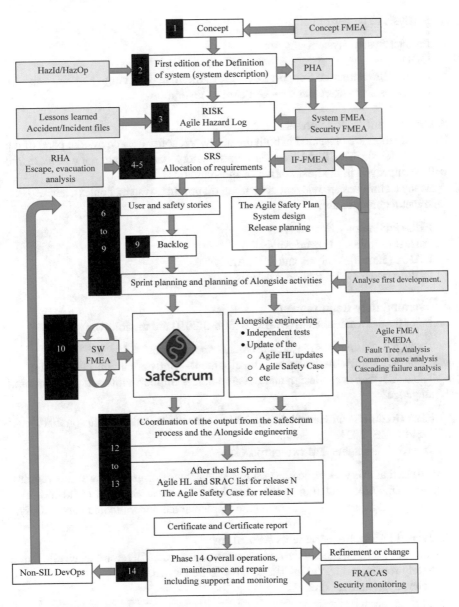

Fig. 4.3 IEC 61508:2010 safety lifecycle together with relevant hazard, risk, and safety analysis

2. Overall Scope Definition Phase

There are no T&M listed in the safety standards relevant for this phase. A good start is to develop the first edition of "Definition of the product/system" and "ODD" description (operational envelope), including relevant scenarios.

Analysis to be performed:

- Product/System/system of systems
- ODD
- ISAD—Infrastructure Support Levels for Automated Driving
- Use cases/scenarios/ConOps—Concept of Operation

3. Hazard and Risk Analysis

Relevant tables are presented in both the hardware, software, and system parts of the safety standards. The T&M listed in current safety standards are not complete and will be improved in the next edition of the standards.

Many techniques are relevant depending on product, system, context, and project. Relevant techniques are:

- PHL (Preliminary Hazard List)
- PHA (Preliminary Hazard Analysis)
- HARA (Hazard Analysis and Risk Analysis)
- LOPA (Layer Of Protection Analysis)

4. Overall Safety Requirements

Some of the relevant tables from IEC 61508-3:2010 are the following:

- Table A.1 "Software safety requirements specification"
- Table A.2 "Software design and development—Software architecture design"
- Table A.3 "Software design and development—Support tools and programming language"

RHA (Requirement Hazard Analysis) is a relevant technique (Ericson 2016) and MIL-Std-882E: 2021.

The RHA technique has two purposes:

- Ensure that every identified hazard has at least one corresponding safety requirement. This link is often included in the hazard log (see Sect. 7.1 of this book).
- Verify that all safety requirements are implemented and validated successfully.

5. Overall Safety Requirements Allocation

No relevant T&M is listed in the current safety standards. Therefore, when allocating the safety requirements, the interfaces have to be known.

Relevant techniques are:

- Interface HazId
- IF-FMEA (Papadopoulos et al. 2001)

Overall Planning: 6. Overall Operation and Maintenance Planning, 7. Overall Safety Validation Planning, and 8. Overall Installation and Commissioning Planning

No relevant T&M is listed in the current standards. See Chap. 2 "Agile Development" in this book. Several approaches are relevant when developing safety-critical products and systems:

- Contingency Planning. The author's experience is that this has often not been sufficiently considered. Developing safety-critical products and systems is challenging, and the possibility that some challenges will surface seems to be greater than what the developers expect when they plan the development.
- Strategic Planning. This has similarities with the "time box" in an agile setting. Strategic planning aims to ensure that the employees and other stakeholders are all working toward a common goal, and that their focus and resources are all aligned towards this.

9. E/E/PE System Safety Requirements Specification
A relevant table from IEC 61508-3:2010 is listed below:

- Table A.1 "Software safety requirements specification"

  Some relevant techniques are:

- RHA updated or refined
- IF-FMEA updated or refined
- Update of the hazard log

10. Realizations
Some of the relevant tables from IEC 61508-3:2010 are listed as follows:

- Table A.2 "Software design and development— Software architecture design"
- Table A.3 "Software design and development—Support tools and programming language"
- Table A.4 "Software design and development—detailed design"
- Table A.5 "Software design and development—Software module testing and integration"
- Table A.6 "Programmable electronics integration (hardware and software)"
- Table A.9 "Software verification"

Realization includes design introduced by software engineers. Some relevant techniques are:

- Agile FMEA
- SW FMEA

Sprint review is an important part of SafeScrum. This occurs at the end of each sprint when the team has produced a coded, tested, and reviewable piece of software (Myklebust et al. 2018). The RAMS manager may be involved in the sprint review. According to Kelly (2020), systematic (design) errors are introduced whenever there is a misalignment between the original intent of a safety requirement and its implementation. Potentially hazardous emergent behaviors may result from well-intended, but in hindsight, flawed design decisions made when addressing or satisfying requirements that, unfortunately, have unintended hazardous side effects. They can also be the result of implementation (process execution) errors during the software development process—e.g., modeling errors, coding errors, and tool-use errors. Therefore, it is necessary to ensure that an assurance effort has been targeted at attempting to reveal both of these sources of errors.

References

Chen, X.H., Snyman, M., Sewdass, N.: Interrelationship between document management, information management, and knowledge management. SA J. Inf. Manag. **7**(3) (2005)

Course Hero. www.coursehero.com/file/50109850/Week-2-Discussiondocx/. Last visited 27 Apr 2021

Ericson II, C.A.: Hazard analysis techniques for system safety, 2nd edn (2016)

Hanssen, G.K., Stålhane, T., Myklebust, T.: SafeScrum – Agile Development of Safety-Critical Software. ISBN 9783319993348. Springer (December 2018)

IECEE Guides. www.iecee.org/documents/refdocs/. Last visited 27 Apr 2021

Kelly. http://dl.acm.org/citation.cfm?id=2894798&CFID=784458267&CFTOKEN=79745997. Project Smart 2014: www.projectsmart.co.uk/. Last visited 11 May 2020

Leveson, N.: A new accident model for engineering safer systems. Saf. Sci. **42**(4), 237–270 (2004)

Leveson, N.: "An STPA Primer". Version 1, 2013 (updated 2015). [Online]. Available: http://sunnyday.mit.edu/STPA-Primer-v0.pdf

Myklebust, T., Lyngby, N., Stålhane, T.: Agile practices when developing safety systems. In: PSAM14 Los Angeles, September 2018

Myklebust, T., Onshus, T., Lindskog, S., Ottermo, M.V., Bodsberg, L.: Data safety, sources, and data flow in the offshore industry. In: ESREL 2021, Angers France

NASA Software Engineering Benchmarking Study NASA/SP-2013-604

NASA Software Safety Guidebook (2004). https://standards.nasa.gov/standard/nasa/nasa-gb-871913

Papadopoulos, Y., McDermid, J.A., Sasse, R., Heiner, G.: Analysis and synthesis of the behaviour of complex programmable electronic systems in conditions of failure. Int. J. Reliab. Eng. Syst. Saf. **71**(3), 229–247 (2001)

Porat, M.U.: The information economy: definition and measurement. United States Department of Commerce, Office of Telecommunications, Washington, DC (1977)

Qatar National Project Management: Document Management Plan Preparation Guidelines www. psa.gov.qa/en/knowledge1/qnpm/Resources/templates/Pages/Phase.aspx. Last visited 25 Apr 2021

SCSC-127E Data Safety Guidance Version 3.3, DSIWG 2021.

SWEBOK. http://swebokwiki.org/Main_Page. Last updated 27 July 2015

Young, G., Leveson, N.G.: STPA: systems theoretic process analysis. An integrated approach to safety and security based on systems theory. Commun. ACM. **57**(2), 31–35 (2014)

Chapter 5
Plans and Functional Safety Management

The best laid schemes o' Mice an' Men
Gang aft agley,
An' lea'e us nought but grief an' pain,
For promis'd joy!

Robert Burns: To a Mouse

What This Chapter Is About
- Safety plans
- Functional safety management
- Software quality assurance plans

5.1 Safety Plan

The (Agile) Safety Plan is a working document, mainly for the safety engineering team (RAMS team). The contents will depend on the project and how the manufacturer organizes the project. For example, in some projects, the SQAP (Software and Quality Assurance Plan) and the safety plan are combined. The safety standards define a safety plan in both the automotive and railway domain but not in the generic standard IEC 61508:2010 or the process industry standards IEC 61511:2016. However, IEC 61511-1:2016 and IEC 61511-2:2016 include a chapter named Safety Planning—A.5.2.4 (Table 5.1).

5.1.1 Safety and Agility

The purpose of the Agile Safety Plan is to ensure that the manufacturer will be specific about the safety process to be used, enabling the internal Safety engineer/ RAMS manager and Certification Body to be proactive and to plan the work according to the development schedule. In addition, the safety plan may also apply

© The Author(s), under exclusive license to Springer Nature Switzerland AG 2021
T. Myklebust, T. Stålhane, *Functional Safety and Proof of Compliance*,
https://doi.org/10.1007/978-3-030-86152-0_5

Table 5.1 Safety planning—automotive vs railway

Automotive	Railway
ISO 26262-1:2018 3.143 safety plan A plan to manage and guide the execution of the safety activities (3.133) of a project, including dates, milestones, tasks, deliverables, responsibilities, and resources.	**EN 50126-1:2017. 3.73 Safety Plan** A documented set of time scheduled activities, resources, and events serving to implement the organization, responsibilities, procedures, activities, capabilities, and resources that together ensure that an item will satisfy given safety requirements relevant to a given contract or project.

for, e.g., the Infrastructure Manager (IM) or the operator. This is because in some projects, both the manufacturer and the IM/operator develop their own, but coordinated, safety plans.

Moving from a waterfall/V-model to an agile model affects several parts of the development process. We have analyzed the EN 50128:2011, IEC 61508-3:2010, and ISO 26262-6:2018 software standards and made an addition to the agile process to ensure that all requirements in the safety standards will be satisfied. The acquired information from safety standards and the agile domain, including the SafeScrum (Hanssen et al. 2018) approach, is used to suggest an Agile Safety Plan that satisfies the requirements in these standards while at the same time enabling an agile development process. The purpose of the Agile Safety Plan is to aid manufacturers in achieving certification or internal approval of their products by satisfying the planning requirements, using the Agile Safety Plan together with other relevant plans.

This Agile safety plan, as described in this chapter, satisfies all requirements mentioned in

- EN 50126-1:2017 chapter 7.3.2.3 "safety plan".
- EN 50129:2018 chapter 5.3.5 "safety plan".
- IEEE std. 1228:1994 (R2002) for software safety plans—reaffirmed 2010, which has been used as a basis for EN 50126–1 safety plan requirements.
- ISO 26262-2:2018 chapter 6.4.6.4 (no title)
- IEC 61508:2010 series
- IEC 61511:2016 series
- IET code of practice for independent safety assessors (ISAs), 2012.

Table 5.2 shows the relevant input documents, related documents, and plans listed.

5.1.2   A Safety Plan and an Agile Safety Plan

Below is a list of all the topics needed for a safety plan together with general information and relevant agile adaptations.

Table 5.2 Input documents and related plans

Input documents	Related plans
• Relevant safety standards, e.g., IEC 61508, ISO 26262, and EN 5012x series • description of the system (DoS) • description of the ODD, including OEDR	**Part of this book** • safety plan issued by suppliers, see Sect. 5.1 in this book • release plan, see Sect. 8.3.1 in this book • SQAP (software quality assurance plan), see Sect. 5.3 in this book • HazOp plans and similar plans if not included in the safety plan
Related documents or work products that are mentioned in safety standards:	
• test plan • verification plan • hardware element evaluation plan • hardware component test plan • safety validation plan • software verification plan • software validation plan • software release and deployment plan • software maintenance plan	**Not part of this book** • Configuration management plan, IEEE Std. 828:2012 and ISO 10007:2017 • Certification plan (Myklebust 2013) • RAM engineering management plan, EN 50126–1:2017 • security plan, see e.g., IET, code of practice: Cyber security for ships. 2017, appendix C "contents of a cyber-security plan—CSP" • Data safety management plan, SCSC Guide 2021

Chapter 1 The policy and strategy for achieving safety should be described.

General comments and Agile adaptations

Policy: A set of ideas or a plan of what to do, in particular requirements that has been agreed to officially by a business organization. For example, in this project, we plan to apply the SafeScrum process.

Strategy: A detailed plan for achieving success in a set of identified situations. For example, this product shall be developed with only the reusable documentation needed to obtain relevant approvals.

Agile adaptations could be performed in all parts of the lifecycle, including a DevOps approach.

Chapter 2 The scope and context of the plan.

General comments and Agile adaptations

This includes:

- Relations to the project plan and the SQAP.
- Description of the product or system to be developed—normally only a reference to a document describing the system.
- Intended use and operational domain (operational envelope)—ODD (Operational Design Domain) and OEDR (Object and Event Detection and Recognition). See, e.g., SAE3016:2019.
- References to relevant safety standards.

A reference to a contract is often included in the scope. The contract could be between a railway operator and a manufacturer or an OEM and a manufacturer.

The agile community "embrace change." We may thus expect changes of the scope of the project several times during an agile project. In other projects, scope changes, including scope creep, should be handled through change control. Normally the safety requirements are more stable than the other requirements (see also chapter 24)

Chapter 3 Constraints and assumptions made in the plan.

General comments and Agile adaptations
Be aware that sometimes constraints and assumptions are mentioned in several places in, e.g., the SC and similar documents. The constraints and assumptions have to be checked by the manufacturer (and others) and, if possible, improved—e.g., gathered in one chapter—in later versions of the relevant documents.

This chapter has a link to the "Introduction" chapter of the safety case.

There is no need for special Agile adaptations.

Chapter 4 Safety organization.

This chapter contains details of roles, responsibilities, competencies, and relationships of distributed teams, suppliers, and bodies (NoBo or CB) undertaking tasks within the lifecycle, including DevOps when relevant.

General comments and Agile adaptations
This part includes definition of roles—see Chapter 3 in the *TASC* (*The Agile Safety Case*) book (Myklebust and Stålhane 2018) for a more complete description of roles:

- Project manager.
- RAMS manager (safety engineer).
- Testers and verifiers.
- Validators.
- QA roles.
- Auditors.
- Assessors.

This part has a link to the "Organizational Structure" chapter of the QMR and the "Safety Organization" chapter of the SMR—see Chapter 6 of the *TASC* book. The authors of the SCs should be identified to ensure good communication between the authors of the SCs and the ISAs.

The Sprint team and relevant safety engineers outside the Sprint team and, e.g., the distributed teams should be defined. The automotive standard includes a definition for ISO 26262-1:2018 3.36 distributed development which goes as follows: *Development of an item or element with development responsibility divided between the customer and supplier(s) for the entire item or element (3.41).*

Note 1 to entry: Customer and supplier are roles of the cooperating parties. See also "Quality Assurance in Scrum Applied to Safety Critical Software" (Hanssen et al. 2016) for details regarding the QA role as part of the Sprint team. The project manager (or scrum master) should preferably consider a similar role even if they do not have an agile approach. Several safety standards are weak when it comes to requirements for distributed development.

The automotive safety standard series ISO 26262 include definitions and requirements that are useful. The relevant definitions can be found in ISO 26262-1:2018. In addition, requirements and guidelines can also be found in ISO 26262-4:2011, ISO 26262-6 SW:2018, and ISO 26262-8:2018.

This chapter has a link to "Organizational Structure" chapter of the QMR and "Safety Organization" chapter of the SMR.

This chapter normally includes references to CVs and how competence is maintained.

Chapter 5 Planning the safety activities: tests, analysis, and verifications.

General comments and Agile adaptations
The different T&M, e.g., from Annex A and B in IEC 61508-3, to be performed should be listed. Alternatively, we need a reference to a document listing the T&Ms. to be used. The software T&M parts are often listed in the SQAP document or in a separate T&M document, see Sect. 4.5 in this book.

Safety analysis, especially related to software, is weakly described in current safety standards. The IEC 61508 committee will improve this in the next edition of IEC 61508-3. See also "The Agile Hazard Log" (Myklebust et al. 2017), the agile FMEA approach (Myklebust et al. 2019a), and Analyse first development (Myklebust et al. 2019b).

There should be a clarification regarding which T&M should be the Sprint team's responsibility and which should be the responsibility of the safety team—alongside engineering in SafeScrum. For the evaluations of the total scope for the safety plan, SQAP and T&M should be described by the manufacturer to ensure that there are no missing requirements.

Chapter 6 A description of the system's lifecycle and the safety tasks to be undertaken within the lifecycle along with any dependencies.

General comments and Agile adaptations
Waterfall models, V-model, and agile methods like SafeScrum are currently used for software development. In the future, we expect that DevOps, DataOps, and ML-Ops will also be used when developing safety-critical systems.

SafeScrum is so far (Hanssen et al. 2018) mainly described for the SW development part, i.e., in phase 10 of IEC 61508. Thus, one may, e.g., include other phases of the waterfall safety lifecycle or V-model that are applied together with the SafeScrum process. See also "The role of CM in Agile development of safety-critical software" (Stålhane and Myklebust 2015) and "Important considerations when applying other models than the waterfall/V-model" (Myklebust et al. 2015). The paper RAMSS for the future (Myklebust et al. 2019a) describes how safety can be combined with RAMS and security, including DevOps.

Chapter 7 Ensuring an appropriate degree of personnel independence in tasks commensurate with the risk of the system, e.g., depending on whether it is a minor safety machine or a railway signaling system.

General comments and Agile adaptations
The required degree of personal independence differs among the domains and depends on how well the manufacturer implements the safety standards. Formally, it depends on, e.g., failure probability and the related consequence. Parts of the risk evaluations can be agile—see "Agile Safety Analysis" (Stålhane and Myklebust 2016). Still, much of the risk analysis work is performed by the manufacturer before the first Sprint.

This chapter has a link to "Organizational Structure" chapter of the QMR and "Safety Organization" chapter of the SMR.

Chapter 8 Hazard identification and analysis.

General comments and Agile adaptations
Hazard identification and analysis are often based on already existing hazard logs—see Sect. 7.2 of this book—from the relevant parties, e.g., the manufacturer, OEM, and the purchasing company, together with a new, project-specific, hazard identification analysis. Unintended hazards can result from implementation errors during the coding or due to wrong/unintended use of software tools. Normally the top hazards are described together with relevant sub-hazards. The core hazard—top hazards—for the ETCS (European Train Control System) for the reference architecture is defined as Subset 091: *Exceeding the safe speed/distance as advised to ETCS.*

This part has a link to "hazard log" chapter of the SMR.

For further information, see "Safety Stories: A New Concept in Agile Development" (Myklebust and Stålhane 2016) regarding safety stories, "The Agile hazard log approach" (Myklebust et al. 2017) and regarding "The Agile FMEA Approach" (Myklebust et al. 2017).

Agile Change Impact Analysis is described in "Agile change impact analysis of safety critical software" (Stålhane et al. 2014), and a template for the report is presented in Myklebust et al. (2014a).

Chapter 9 On-going risk management, including dynamic risks.

General comments and Agile adaptations
Ongoing risk management is project-, product-, and context-dependent. Project risk and analysis to identify risks and opportunities (e.g., improved functions) are assumed to be taken care of as part of the project plan. Using existing generic and domain-specific information, it is possible to get an early start on safety analysis. This is important since architectural decisions made early in a project—agile or not—are expensive to change later. For example, FMEA (Failure Mode and Effect Analysis) and its variants IF-FMEA (Interface Focused FMEA) (Papadopoulos et al. 2001) work well in an agile setting, "Agile safety analysis" (Stålhane and Myklebust 2016) and "Agile change impact analysis of safety critical software" (Stålhane et al. 2014).

Cars operate in a highly dynamic environment, which contains frequent changes of hazards, risks, vulnerabilities, and technologies, while involving variable missions and functions. To the author's knowledge, BSI PAS 1881:2020 and UL4600:2020 are the only safety standards that include relevant requirements related to dynamical risk, e.g., "Data Catalog Identity and Access Management."

This part links to the "hazard log" chapter of the SMR and other chapters depending on the risk management results. Based on the results of a CIA, relevant T&M may have to be repeated by the manufacturer.

Chapter 10 Risk tolerability criteria.

General comments and Agile adaptations
This topic is domain and culturally dependent. The criteria to be evaluated are the probabilities for an accident and the resulting consequence. The relevant stakeholder normally defines the safety requirements for a railway signaling system to be SIL4. In the Oil and Gas industry, the NOG 070:2018 lists relevant SILs for different products and systems. The recommended SIL for most of the relevant products and systems is SIL2. Typical automotive systems, such as airbags, are ASIL-D, and radar cruise control systems are ASIL-C.

Chapter 11 The establishment and ongoing review of the adequacy of the safety requirements.

General comments and Agile adaptations
The understanding of the existing requirements may change during the project. The adequacy review may also be dependent on the contract between the manufacturer and the purchasing company. When checking the adequacy of the safety require-ments, the corresponding hazard log items linked to the requirements should be evaluated simultaneously. This may be performed as part of, e.g., the "Backlog Refinement Meeting" also named "Backlog grooming."

Chapter 12 System design.

General comments and Agile adaptations
The design phase should cover the following topics: architecture description, soft-ware requirements specification, hardware requirements specification, and a test plan that includes the relevant tests and analysis for the type test to obtain official approval of vehicles before they can be put on the market.

When the system is based on products that already have a corresponding safety case and a safety manual, these documents have to be studied carefully and taken into account when performing, e.g., tests and analysis.

Regarding API, see chapter 24 of the safety plan below.

According to experience: www.weibull.com/hotwire/issue186/fmeacorner186. htm

In my experience, 50% or more of system problems occur at the interfaces between sub-systems or components, or as a result of integration with adjacent systems. Understanding and addressing interfaces and integration is essential to achieving safe and reliable systems. A System FMEA is uniquely capable of making interfaces and integration issues visible, and addressing them through the FMEA procedure.

Relevant methods to be used are, e.g., FTA—see IEC 615025:2006—and Markov analysis—see IEC 61165:2006. This is part of the chapter "System Design" or "System/Sub-system/Equipment Design" in the SMR part of the SC. For incremental design development, we also require a thorough configuration management plan. See, e.g., "The role of CM in Agile development of safety-critical software" (Stålhane and Myklebust 2016).

Chapter 13 Tests, verification, validation, and regression.

General comments and Agile adaptations

A reference to the test plans and V&V plans should be included. Regression validation and Regression testing have become more important since signaling systems now include more SW that are easier to update and upgrade than HW and due to the increased use of modern SW development methods.

Current safety standards are weak when it comes to regression testing requirements. Being able to repeat tests frequently and without too much additional costs to reassure that all relevant parts of previously checked code still operate as intended—defined by tests—is an ideal to strive for. Regression testing has two benefits: first, it creates confidence with developers that the system operates as intended, that recent changes do not make previously checked code fail, and that it is OK to move on. Second, it also creates confidence with other stakeholders, e.g., the customer that the system performs as intended.

With the recent adaptation of agile methods such as SafeScrum to the development of safety-critical systems and efficient tools for test automation, it is possible to do regression testing of larger parts of the system, with increased frequency, without adding extra cost. First, practices such as test-first development enable testers to continuously develop the test-suite alongside—in parallel with—creating code—somehow reducing the need for dedicated testers. Second, by using tools (e.g. testing and analysis tools), testing, and integration frameworks, tests can be repeated more often, e.g., through a nightly build regime. In this way, all potential conflicts or errors that recent changes to the code-base have caused will be known shortly after they have been introduced, and can thus be resolved when the knowledge about the code is fresh in mind. The problems are thus easier to resolve.

To summarize, an emphasis on building the test suite alongside the code, in combination with tools for test automation, will enable regression testing without imposing excessive extra costs. If the number of tests grows too large to be run effectively frequently, e.g., every night, we may use an approach where we select the "right" parts of the system to re-test. Beck and Andres (2004) suggests repeating tests that have previously failed under the assumption that they are more likely to identify problems. Regarding integration, see ERA "clarification note on safe integration" (ERA 2020).

This part has link to "V&V" chapter of the SMR.

In SafeScrum projects, one should specify which parts of the tests and verifications are performed as part of the Sprints and which parts shall be performed by the "Alongside engineering" team. Current safety standards are weak when it comes to

ML/AI requirements. Further research on V&V approaches has to be performed before ML and AI relevant methods are used in the development process.

Chapter 14 Safety audit, to achieve management process compliance with the safety plan.

General comments and Agile adaptations
We need to describe the requirements for periodic safety audits and safety reviews throughout the lifecycle appropriate to the safety relevance of the system under consideration, including any personnel independence requirements. However, the safety standards are not specific concerning how often safety audits should be performed and how rigorous the safety audits should be. ISO 19011:2018 "Guidelines for auditing management systems" is of help when planning and performing safety audits.

This has a link to "safety review" chapter of the SMR.

CLC/TR 50506-2:2009 states: *If planned, external and/or internal safety audits can be held in order to analyze the safety management, then these audits should be documented in compliance with the relevant safety standards.* This has a link to the chapter *"safety reviews"* of the SMR. It is important to establish a strategy for the safety reviews, see e.g., DNVGL-RP-O101 (2016). For example, Identify documentation for review from a risk and criticality perspective in order to reduce the number of documents requiring multiple revisions. Safety audits performed by the assessor should be part of the planned communication between the assessor and the supplier.

Chapter 15 Documentation, information, and work products (automotive).

General comments and Agile adaptations
See Annex A of this book, which includes an overview of relevant PoC documents mentioned in safety standards. The latest edition of ISO 9001:2015 is goal based when it comes to documentation. For example, one of the most important objectives in the revision 2015 is the amount and detail of documentation required in order to be relevant to the desired results of the organization's process activities. ISO 9000:2015 clause 3.8.5 gives the following examples of allowed documentation: paper, magnetic, electronic or optical computer disc, photograph, and master sample.

Large projects must have a separate documentation management plan. Discuss with the assessor which documents are needed and what, e.g., need only to be information as part of databases and tool logs. See "Scrum, documentation and the IEC 61508-3:2010 software standard" (Myklebust et al. 2014b).

Chapter 16   Hardware.
Hardware can be developed using an Agile approach, e.g., by using FPGA, for further information, see, e.g., the blog at Infoq (2015).

General comments and Agile adaptations
Hardware components can be split into two major parts: components with inherent physical Properties—see EN 50129:2018, C.4, and programmable components or devices. If HW development is included, an EMC (electro magnetic compatibility) compliance plan should be referenced to or included in this plan. EMC is important

and difficult to satisfy for the manufacturers that develop HW. One of the authors has experienced, since the issue of the EMC directive 89/336/EEC in 1989, that more than 90% of the products fail in at least one test the first time EMC tests are performed. The EMC directive came into force in 1992 and has been mandatory for CE marking of electronic/electrical products since January 1, 1996.

This section has a link to "assurance of correct hardware functionality" chapter of the TSR. Note that while software can include changes, e.g., once a week, hardware is not changed more than, e.g., once a year.

Chapter 17 Software.

General comments and Agile adaptations

The main parts of the software plan are described in the SQAP, see Sect. 5.3 of this book. However, one of the authors has experience in several projects within the railway domain that the required "software assessment report" is often not mentioned.

This has a link to "assurance of correct software functionality" chapter of the TSR. When using agile methods, see the SafeScrum book (Hanssen et al. 2018).

Chapter 18 Data.

General comments and Agile adaptations

Due to the increasing digitalization of safety systems, data has become far more important. Relevant data are

- Configuration data: For example, railway control systems and air traffic control systems typically have large amounts of data to adapt the generic software to the given control area.
- Geographical data: Similar to the example above but derived from external sources (e.g., nautical chart data). Used in, e.g., autonomous and semiautonomous vehicles.
- Datasets imported automatically or entered manually. For example, health informatics data.
- Real-time data derived from sensors. This is normally part of the SRS and as such addresses indirectly by IEC 61508:2010.

If data is of importance for the development project, the SCSC Data Guide (2021) should be adapted for the project.

The railway standard EN 50128:2011 includes a chapter named "Development of application data or algorithms: systems configured by application data or algorithms" that should be applied for the railway domain and could preferably be a basis for other domains. The standard does also mention a few data techniques in Annex D "Bibliography of techniques," e.g.

- Data Flow Analysis.
- D.11 Data Flow Diagrams.
- D.12 Data Recording and Analysis.

The automotive ISO 26262-6:2018 includes a few requirements related to data as e.g.:

- The software architectural design shall describe:

  (a) The static design aspects of the software architectural elements; the data types and their characteristics.
  (b) The dynamic design aspects of the software architectural elements; the data flow through interfaces and global variables.

- Table 4, table 7 and table 10 in the standard including data flow analysis.
- Configuration data.
- Calibration data.
- Design principles for software unit design and implementation at the source code level as listed in Table 6 (in the standard) shall be applied to achieve the following properties:

  (a) Correctness of data flow and control flow between and within the software units.

- Table 6 including no hidden data flow or control flow.

Changing data is easier than changing software, so agile thinking could be helpful. This section has a link to "assurance of correct data" chapter of the TSR.

Chapter 19 A process to prepare relevant safety cases: if relevant.

General comments and agile adaptations
Whether a safety case shall be developed depends on the domain—e.g., it is required by the railway domain—and the relevant regions the product or system shall be taken into use. A process to prepare the Safety Manuals could be similar to the manuals described in IEC 61508:2010 series and should preferably be coordinated with the safety case preparation. Both ISO 26262 and UL4600 require a safety case, but these standards are not required to be used by law.

This information may include, scope and structure of the safety case, principal components, safety case authors, and timescale for the delivery of the safety case references and the safety case itself. In the railway domain, some information related to ISA deliverables and safety authorities are usually presented together with e.g., a diagram showing the documents resulting in a SASC (Specific Application Safety Case).

In the "Related safety case" chapter in the SC, the SARs, issued by the assessors, shall be referenced. The safety case should preferably be developed incrementally alongside the Sprints.

Chapter 20 A process for safety approval of the system.

General comments and agile adaptations
The approval process will vary between the domains. The safety approval has a link to the "system/sub-system/equipment handover" chapter of the SMR. As always,

communication with all relevant stakeholders is important. The process for system safety approval should be evaluated together with the deployment/release plan.

Chapter 21 A process for the safety approval of modifications.

This process should be evaluated together with the deployment/release plan.

General comments and agile adaptations

Before any modification, the evidence must be provided to show that the modifications will not adversely affect the safety properties of the unmodified rest of the system. This work should be done by developing a change impact analysis (CIA). Agile CIA is described in "Agile change impact analysis of safety critical software" (Stålhane et al. 2014). This should also apply to patches. The IEEE std. 24,765:2010 defines a patch as a modification made to a source program as a last-minute fix. Other standards like the IEC 62443 security series have other definitions for patching.

Chapter 22 A process for analyzing, operation, and maintenance performance to ensure that realized safety is compliant with requirements.

General comments and Agile adaptations

Some companies use a FRACAS (Failure Reporting, Analysis, and Corrective Action System) approach (MIL-HDBK-2155 2014). For the railway domain, see, e.g., information related to "safety qualification tests" in chapter 5.3.12 EN 50129:2018. BSI PAS 1881:2020 will specify how to assure the safety of automated vehicle trials and testing.

We foresee that having an agile approach will make it more convenient to update and upgrade the software if planned or necessary due to, e.g., security challenges. In a DevOps perspective, this becomes important also due to the possible improved monitoring possibilities.

Chapter 23 A process for maintaining safety-related documentation—information and working products, including a Hazard Log.

General comments and Agile adaptations

See Annex A of this book, which includes an overview of relevant PoC documents mentioned in safety standards. In small projects, the relevant documents are listed in the safety plan. For further information, see e.g., " SECTION 2 DOCUMENT MANAGEMENT—BEST PRACTICE" in "Recommended practice: Technical documentation for subsea projects" (DNVGL-RP-O101, 2016) and the Agile Hazard Log approach (Myklebust et al. 2017).

Chapter 24 Subcontractor management arrangements, safety assessment, to achieve compliance between system requirements and realization including subsystem and the system.

General comments and Agile adaptations

Subcontractor management arrangements are linked to the "system/subsystem/ equipment design" chapter of the SMR. Normally, a subcontractor contract is used as a basis.

Integrate in practice:

In computer programming, an application program interface (API) is a set of subroutine definitions and communication protocols. An API describes how to interact with a separate software component or resource. In general terms, it is a set of clearly defined methods of communication between two or more components, items, or products. A good API should make it easier to develop a computer program by providing all the building blocks, which the programmer then puts together. An API specification can take many forms but usually includes specifications for routines, data structures, object classes, variables, or remote calls. POSIX, Windows API, and ASPI are examples of APIs. Documentation for an API usually is provided to facilitate usage and implementation. However, it is important to remember that API abuses are set to become the most frequent attack route for data breaches by 2022. In times of uncertainty, it's critical to protect your valuable data by adding layers of security through APIs. Security in any workforce environment is becoming top of mind for IT and OT leaders who are prioritizing data security.

Safety issues:

The 2010 edition of IEC 61508 introduced the concept safety manual. The purpose of a safety manual for a compliant item is to document all information relating to the item. This is required to enable the integration of the compliant item into a safety-related system, a subsystem, or an element, to comply with the requirements of this standard. The safety manual is not mentioned in the CENELEC EN 5012x standards. However, it is still important, especially at the GP level since designers and integrators of products, equipment, or systems need the information presented in the safety manual to ensure that the integration can be performed without compromising safety. Requirements for the content of the safety manuals are presented both in IEC 61508-2:2010—Hardware part of the requirements—and IEC 61508-3:2010—Software part of the requirements.

Contract and safety issues, including customer-supplier interacting:

A related document to API and safety manual is the DIA (Development interface agreement) in the automotive safety standard ISO 26262. The safety planning is documented and references the development interface agreements (ISO 26262-8:2018, Clause 5) that define the interfaces with the safety plans of the other parties. Annex B "Development Interface Agreement example" presents an example of a DIA. Relevant topics in a DIA are:

- Prequalifications.
- Qualifications.
- Acceptance of conditions.
- Capability issues.
- Safety goals.
- Compliance issues.
- Iterative evaluations.
- Safety plan and HARA.
- Proven in use.
- Development lifecycle issues.
- Design issues.
- Integration issues.
- Tests, verifications, and validations.

The Norwegian Agency for Public Management and eGovernment has issued guidelines for agile contracts. For further information, see www.anskaffelser.no/verktoy/smidigavtalen-ssa-s

5.1.3   Summary

An Agile Safety Plan ensures a good start of the development project, minimizes costs, and reduces time to market. It also ensures that the safety process is complete and produces sufficient information to be developed by the manufacturer and reviewed by the certification body.

Suggestions for improvements of current safety standards are:

- IEC 61508 should include requirements for a safety plan and safety case. The safety plan requirements could be similar to EN 50126-1:2017 ch. 7.3.2.3 requirements, and the SC requirements could be similar to the requirements for an SC in ISO 26262-2:2011.
- EN 50128:2011 should include requirements for a safety manual. The safety manual requirements could be similar to the IEC 61508:2010 requirements for a safety manual.
- Improved release planning, especially when having an agile approach.
- IEC 61508-3 should improve the deployment part, e.g., based on current EN 50128:2011.
- Guidelines for distributed development, although the ISO 26262 series includes some requirements for distributed development.

5.2   Functional Safety Management (FSM)

5.2.1   Introduction

We will start by quoting S. Gandy (Gandy 2017): "FSM is designed to ensure that ALL stages of the safety life cycle are properly implemented and supported, in terms of—what I like to call—the three Ps: People, Paperwork and Procedures. The people aspect relates to the roles, responsibilities and competency of personnel involved in SLC activities; the paperwork relates to documentation and record keeping and the procedures relate to having well-defined work processes in place for each phase of the SLC: Analysis, Design/Implementation and Operation/Maintenance." We will add personnel experience and information to the list of what is included in the three Ps.

In practice, the functional safety management usually is a two-step job, using the safety plan, based on the safety plan template—see Sect. 3.1—as a starting point. Before the development starts, we need to see if the project will be able to meet all the requirements of the safety plan. When the project is finished, we need to check that they have done all safety-related activities according to the safety plan—the same way as we do general quality assurance.

In addition, it is important to ensure that the company has a safety culture. If possible, this should also be checked periodically. There are two types of safety culture and their relevance depends on what they do. For a production company— e.g., a factory—the safety culture shall keep the employees safe. For a development company—e.g., a software company—the safety culture is about making a safe product.

5.2.2 Safety Culture

There are several checklists on safety culture available. We have chosen a short one for this section, used by the International Atomic Energy Agency (IAEA) (see Wolniak and Olkiewicz 2019). Our comments are added in italics.

1. Commitment to safety and safety culture is needed. *Important but difficult to verify.*
2. All applicable procedures shall be used. *The procedures need to leave a paper trail, not necessarily documents.*
3. Conservative decisions shall be taken. Use the STAR principle: Stop-Think-Act-Review. *Important principle but difficult to verify.*
4. Near misses and failures shall be reported. *An open-minded and transparent culture.*
5. All unsafe factors and conditions shall be identified. *An open-minded and transparent culture, learning from e.g., service agreements.*
6. Safety and quality shall be improved continuously. *This is in line with the ISO 9001.*
7. Responsibilities and interfaces shall be known. *This is the management's responsibility. For autonomous teams, the team will also be responsible.*

See also ISO (2015). In addition, the checklist in Dick et al. (2004, Appendix C) will be useful. An example is shown in Fig. 5.1. Using this and other related checklists will give a good idea of how each employee adapts to the safety culture.

QUESTION 2 – WHAT WE'RE DOING		Don't Know	Strongly Disagree	Disagree	Maybe	Agree	Strongly Agree
2a	I think that the programme is aware of data safety risks.	☐	☐	☐	☐	☐	☐
2b	I believe we need to implement measures to manage data safety risks.	☐	☐	☐	☐	☐	☐
2c	I think that the programme meets its obligations (e.g. has a Data Management Plan in place and a role with specific responsibilities in this area).	☐	☐	☐	☐	☐	☐

Fig. 5.1 Checklist example copied from the SCSC Data Safety Guidance, © SCSC

5.2.3 Safety Plan Related Issues

Below we have included all the 24 issues mentioned in the safety plan template—see Sect. 3.1 "Safety Plan and safety Plan Template"—and discussed how we could check the project's ability to perform and the quality of the actual performance of each issue. At first glance, it might look like a large amount of paperwork. However, a lot of the paperwork is already part of the standard quality assurance (QA) process. If the QA process is under control, it will later suffice with spot-checks. A good planning process including feedback from other projects will also help to build confidence in the planning process and that it is followed up internally in each project.

Chapter 1: The policy and strategy for achieving safety should be described.

We need a reference to the company or project's policy and strategy. The policy and the strategy must be considered to be in line with the company's resources and the project's goals and requirements.

Chapter 2: The scope and context of the plan.

Before the project starts, we need to check the safety requirements and the intended environment—Operational Design Domain (ODD)—to see if they are in agreement with the risk assessment. We also need to check that the relevant standards have been used in the planning process and that all plans are in accordance with these standards.

When the project is finished, we need to check the paper trail/information to see if everything is done according to the plan and to the relevant standards.

Chapter 3: Constraints and assumptions made in the plan.

Constraints and assumptions are important since they strongly influence the safety plan and any other plan. In addition, they might be included in the system's manuals. Two things must be checked:

- Are the plans in accordance with the constraints and assumptions?
- Are plans changed or checked for changes needed when something changes and makes constraints or assumptions void. This implies that somebody must be responsible for keeping watch on relevant changes.

Chapter 4: Safety organization.

To check this, we need answers to one question: Are all planned positions filled with people with the right competence and experience? We might also need to check if relevant standards' requirements of independence are fulfilled in the project's organization.

Chapter 5: Planning the safety activities: tests, analysis, and verifications.

As for the other activities, three things are important. Are the planned activities:

- In accordance with the safety requirements, the safety plan, and all relevant standards?

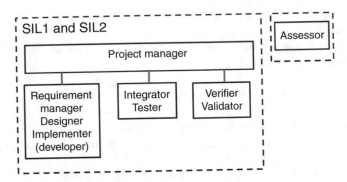

Fig. 5.2 Roles and independence in EN 50128, SIL 1 and 2

- Staffed with competent and experienced people?
- In accordance with the intended use?

Chapter 6: The system's lifecycle and relevant safety tasks.

The software development lifecycle has a set of activities that are to be performed in series or in parallel. The safety-related activities that are needed are defined in the safety plan. The FSM must check that all activities needed are included in the lifecycle document ad that they will leave a paper trail. When the project is finished, it is necessary to check that all the included safety activities have been performed according to relevant standards.

Chapter 7: Ensuring the necessary degree of personnel independence.

Some standards have requirements regarding the independence of people doing, e.g., safety analysis or system testing. The degree of independence must, to a certain degree, be commensurate with the consequences of errors in operation. In addition, several standards have requirements regarding independence, depending on the SIL number. An example from EN 50128 is shown in Fig. 5.2.

Roles that are enclosed with a dotted line can be in the same organization. Roles that are enclosed with a solid line may be the same person.

Chapter 8: Hazard identification and analysis.

There are two things that need to be managed and checked for hazard identification and analysis—the methods and the information used. The methods should be identified at the start of the project. The information used will, however, vary over time.

The first hazard analysis will have to be based on the high-level design, information about the planned operational environment plus any information available from the hazard log and from relevant generic hazard lists.

The functional safety management responsible needs to check that

- The methods used are appropriate for the system and within the standards relevant for the identified application domain.

- The persons who do the hazard identification and analysis have the right competence.
- That all information used is up-to-date and relevant for the system and intended operational environment.

Chapter 9: On-going risk management.

In a static world, risk could be assessed once—at the start of the project. However, in most projects, many things might change over time, e.g., requirements, operating environment, design, and solutions used to achieve the required functionality. Thus, the risk might change.

The project needs to have the following mechanisms in place:

- Keep an overview over all assumptions related to system-related risks.
- Check the assumption list periodically to see if there are changes that will influence the risk assessment. How frequent the assumption list should be checked must be agreed between the project and the assessor.
- If an assumption changes, the parts of the risk assessment influenced by the changed assumption must be repeated. If this gives a new risk level—e.g., SIL—it might be necessary to change one or more project activities.

Chapter 10: Risk tolerability criteria.

This chapter is strongly related to the previous one—ongoing risk management.

An application should define and maintain a risk tolerability level. The simple way to do this is to define maximum levels for each parameter used to assess the system's risk—e.g., for IEC 61508, this will include event consequences, frequency, possibility to avoid the consequences, and the probability of the event. In some areas, the authorities have defined the required risk level—e.g., for railway, all signaling systems should be developed according to SIL 4 requirements.

There are, however, also several other ways to achieve this—see for instance Sect. 5.4 on ALARP and GALE.

Chapter 11: The establishment and ongoing review of the safety requirements.

This chapter is strongly related to chapter 9—Ongoing risk assessment. As in many others of the activities related to FSM, this chapter also has two sets of activities—one before the implementation and one after.

Before development starts, we must check that the safety requirements are based on solid knowledge of the system's functionality and its operational environment—see also chapter 8—Hazard identification and analysis. The safety requirements must be considered enough to avoid or mitigate all identified hazards. The ongoing risk assessment will alert us if one or more of the safety requirements need to be changed.

Both chapters 8 and 9 leave a paper trail. Thus, in order to check if the safety requirements have been under control, we need to check that the activities described in chapters 8 and 9 have been performed satisfactorily.

Chapter 12: System design.

According to the safety plan, system design includes the following documents: architecture description, software requirements specification, hardware requirements specification, and a test plan. The functional safety management must check that these documents are available and are reviewed and agreed upon by the project. When the project is finished, it is important to check that the system is made according to the relevant documents—architecture description, software requirements description, and so on. We recommend that the documents from chapter 1: a review of safety requirements to see if all relevant documents are updated if the assessed risks are changed.

Chapter 13: Tests, verification, validation, and regression.

First and foremost—the project must have a test plan. See chapter 8—Test, analysis, and V&V. The task of the FSM personnel is to check that these plans are in accordance with the system's requirements. When the project is finished, it is important to check that test, analysis, and V&V are done as specified and that the results are as expected.

Chapter 14: Safety audit.

The purpose of the safety audit is to check management process compliance with the safety plan. Before the project starts, we need to see if the planned management process is compliant with the safety plan. When the project is finished, we need to see if the management process ran as planned. Thus, it needs to leave an auditable paper trail.

Chapter 15: Documentation, information, and work products.

Documentation is important for all products. It is needed, e.g., for installation, use, and maintenance. The safety plan must contain a list of documentation, information, and work products needed. When the project is finished, we must check that all documentation needed is ready for use.

Chapter 16: Hardware.

Much hardware is standard components and does not need to be included in the safety plan. Some hardware is, however, custom made or customer adapted, e.g., using an FPGA (Field Programmable Gate Array) or standard hardware with embedded software. Development and validation of such hardware need to be included in the safety plan. The same holds for any updates—planned or not planned.

Chapter 17: Software

Most aspects of software safety are covered in the other suggested chapters. What is not there is the software quality assurance. The project needs to have a plan for how to do quality assurance according to the required quality assurance standard—e.g., ISO 9001. When the project is finished, the quality responsible must check that all required activities have been performed by checking their "paper trail." See also Sect. 3.3 Software Quality Assurance Plan.

Chapter 18: Data

In safety-critical systems, data are used in several ways; as configuration information, input to and output from algorithms—e.g., input from sensors and resulting output to operators and actuators and data used to calibrate instruments. Data will in some cases also have a value as part of a machine learning process—MLOps, using the same ideas as DevOps (Merritt 2020). In addition, safety-critical systems may be used to store important data—e.g., personal information. As a consequence of this, the safety plan must contain information on how to protect critical data.

Data will be changed, either because the environment changes or because one or more applications are changed. Thus, we need a process for how to change data in a safe way.

When the project is finished, it is necessary to check that all safety measures are satisfied.

Chapter 19: A process to prepare relevant safety cases.

We recommend that a development project starts working on the safety case as soon as possible. The safety plan needs to contain important things such as the "language" used for the safety case—i.e., prose, structured prose, or a graphical notation. The safety plan should preferably include a plan for safety case updates. Leaving it until the end of a project is usually a bad idea.

When the project is finished, we need to check that the safety case is developed according to the safety plan.

Chapter 20: A process for safety approval of the system.

All safety-critical systems will have a process for safety approval. This plan should be reviewed at the start of the system. It is important that issues such as the user manual, assumptions for environment, and hardware are properly described. This should include how the process will leave a "paper trail" for later verification.

When the system is finished and approved, we need to check that the earlier reviewed approval process has been followed.

Chapter 21: A process for the safety approval of modifications.

Most software systems are modified one or more times during their lifetime. This tendency will increase with an increasing focus on agile development and DevOps. How to take care of safety during software updates must be part of the safety plan. Two process steps are important and need to be described:

- Change impact analysis—What are the consequences of the modification?
- Regression testing—Is the system still behaving as intended? The FAT will be useful here.

Chapter 22: A process for analyzing operation and maintenance performance.

This activity is needed to ensure that the realized safety is compliant with the requirements. This chapter is closely related to chapter 20—A process for safety approval of the system.

Performance is linked to some important system characteristics—e.g., needed memory space, network capacity, and required response time. The required process must contain descriptions of how to assess these characteristics.

Chapter 23:   A process for the maintenance of safety-related documentation.

Safety-related documentation consists of documented safety reviews, reviews of safety, and hazard analysis plus all relevant V&V activities. In addition, it includes the Hazard Log—how it was used and how it is updated. A process for keeping this under control must be in place before the project starts. At the end of the project, it must be possible to see that the job is done properly. This holds also for updating the hazard log.

Chapter 24:   Subcontractor management arrangements.

This activity contains control of safety assessment and safety-related activities done by the subcontractors needed to achieve compliance between system requirements and realization. It is important that the same control mechanisms and safety requirements are the same for the development company and its subcontractors.

5.3   Software Quality Assurance Plan (SQAP)

5.3.1   Introduction

The main structure of this section is based on a small booklet called "ISO 9001 for Small Businesses" (ISO 2015). In addition, we have included ideas and requirements from ISO 26262—automotive, IEC 61508—generic, and EN 50128—railway. Quality assurance is important because it will help us to

- Keep our promises to the customer—e.g., we promise to review all developed code, and the QA system will make sure we do.
- Let all required activities leave a paper trail.
- Improve the way we do our jobs—i.e., identify our mistakes and find a way to rectify them.
- Be compliant with relevant functional safety standards and public rules and regulations.

In order to achieve all this, all required activities have to leave a paper trial as this is important since it is needed both for proof of compliance—we need to prove that we have done what we said we would do—and for the construction of a safety case. If you want to claim compliance with one of the standards, you will need an audit from a certified assessor and a written confirmation of compliance.

Agile development will give us both opportunities and challenges that are not present when using other development methods. Thus, we have discussed agile development in a separate section. Other life cycles models will be discussed if they cause special needs—e.g., other types of process data should be collected.

Together with the software quality assurance plan, it is important to consider some related plans, e.g., the safety plan and a discussion of tools and methods—T&M (Sect. 3.1), the safety requirements specification—SRS (Sect. 4.2), and tools validation (Sect. 7.1).

5.3.2   *Quality Standards: The Important Parts*

This section will use ISO 9001 as its starting point. Requirements from other relevant standards will be discussed in connection to the relevant ISO 9001requiremnts. We need to consider five issues: organizational requirements, quality culture, product requirements, review of product requirements, and communication. Each of these issues will be presented below.

5.3.2.1   Organizational Requirements

ISO 9001 states that the first step in building a quality assurance (QA) system is to understand the organization and its context and the needs and expectations of interested parties. In addition, we need to determine the scope of the quality management system and its processes. The company's management must show leadership and commitment—first and foremost, they need to have customer focus and establish a quality policy which must be communicated. In order to make the QA

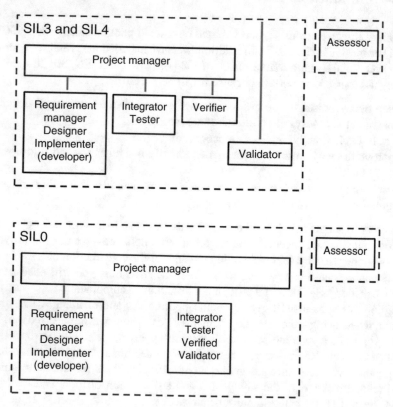

Fig. 5.3 Independence requirements according to EN 50128 for SIL 0, SIL 3 and SIL 4

system work, the company also needs to make clear the organizational roles, responsibilities, and authorities. Independences between some of the roles are also important—see diagram below. Independence requirements will vary depending on the SIL value and the application domain. Roles included in a dotted line can be in the same organization (Fig. 5.3).

IEC 61508 only states that the organization shall have an appropriate quality management system, while ISO 26262-2 states that the QA management system shall comply with a quality management standard, such as International Automotive Task Force (IATF) 16,949 in conjunction with ISO 9001, or equivalent. The same holds for EN 50128.

5.3.2.2 Quality Culture

Top management shall demonstrate leadership and commitment with respect to the quality management system. Thus, they need to make sure that the QA management system is an integrated part of organization's business process, that the necessary resources are available, and that the QA is communicated to all relevant personnel. ISO 26262, IEC 61508, and EN 50128 do not discuss this issue. ISO 26262 has requirements related to safety culture but quality culture is not mentioned.

Note that safety culture is different from quality culture (see Wolniak and Olkiewicz 2019). At the top level, the author provides the following relationship: "safety culture" + "quality culture" = "integrated management system culture." The paper also contains a list of features characterizing the two types of culture. The following is just a small example.

- Quality culture: Commitment of management and employees to all kinds of activities aimed at improving and maintaining quality within all activities in the organization.
- Safety culture: Involvement at the managerial level requires emphasizing on the priority of security, and on the common goals between managers and employees.

5.3.3 Product Requirements

The organization shall collect and understand the requirements specified by the customer. In order to decide the system or service requirement, we first of all need to communicate with the customer. Note that there are requirements that are expected even if they are not stated—implicit requirements. Consequently, we must involve personnel that know the operational and business environment of the product.

The IEC 61508 and the ISO 26262 are only concerned with safety requirements. EN 50128 refers to ISO 9126 for product requirements and quality. ISO 9126, "annex A provides recommendations and requirements for software product metrics and quality in use metrics… These metrics are applicable when specifying the quality requirements and the design goals for software products, including

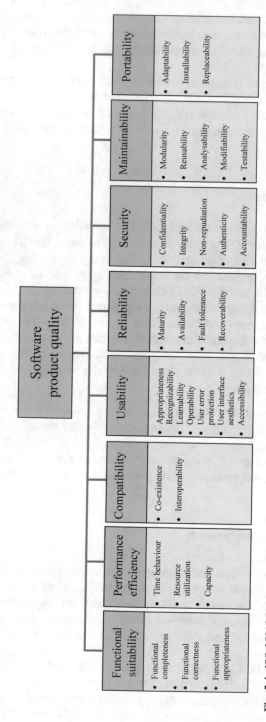

Fig. 5.4 ISO 25010 Product quality model. The model from ISO/IEC 25010:2011 is reproduced by SINTEF and NTNU in the Springer publication Proof of compliance under licence from Standard Online AS June 2021. © All rights are reserved. Standard Online makes no guarantees or warranties as to the correctness of the reproduction

intermediate products." However, according to ISO 9126, safety—as opposed to reliability—is not a quality attribute. ISO 9126l has now been replaced by ISO 25010: 2011. This new standard also contains an annex A.1, which identifies the differences between ISO 9126-1 and ISO 25010. In Fig. 5.4, we list relevant system and software product quality aspects.

When the company develops a product for the marketplace, there is no customer to communicate with. In this case, the company needs another person to take the customer role—e.g., someone from the marketing department or one or more agents who sell their products.

5.3.3.1 Review of System and Product Requirements

When the requirements are documented, they must be reviewed, first and foremost, to see if the organization can meet the requirements. If the requirements are changed, the review must be repeated. Make sure to include all relevant documents—also the review of whether the company has the ability to deliver according to the new requirements.

IEC 61508 requires that the safety requirements shall be reviewed for completeness, correctness, freedom from intrinsic specification faults, understandability, freedom from adverse interference of non-safety functions, and providing a basis for verification and validation. In IEC 61508-1, 7.7.2 the standard requires that users shall be given the information needed to enable the user to ensure the required functional safety during operation and maintenance. This could, e.g., be relevant SRAC's, etc.

ISO 26262-2 takes review a step further by providing an extensive table—Table 1: "Required confirmation measures, including the required level of independence," while EN 50128, Table B7 states that "the validator shall review the software requirements against the intended environment / use." These are just examples. ISO 26262-2, Table 1 has 13 entries, and EN 50128, Table B7 has 17 entries. Thus, the relevant standards should be consulted by the relevant actors when needed.

5.3.3.2 Customer Communication

Quality assurance is first and foremost about satisfying the customer's explicit and implicit requirements. Explicit requirements require good communication with the customer while implicit requirements require an understanding of the customer's market. Thus, customer communication is the first issue to be considered in quality assurance. "Popular communication channels during safety analysis include formal meetings, project coordination tools, documentation and telephone." Wang et al. has done research on the use of communication channels in the development of safety critical software. Their conclusion is that: "Email, personal discussion, training, internal communication software and boards are also in use" (Wang et al. 2018).

IEC 61508-7 mention communication with the customers in section C.2.1.2 where it states that we need to ensure that all the requirements are determined and

Table 5.3 The CORE viewpoint hierarchy and a small example taken from (Stålhane and Malm 2020)

Source The viewpoint from which the input comes	Input The name of the input item	Action The action performed on one or more inputs to generate required outputs	Output The name(s) of any outputs generated by the action	Destination The viewpoints to which the output is sent
Example Gate sensor	The gate is opened	Send "gate opened" to central unit	Send "stop robot" to tool cell	Robot tool

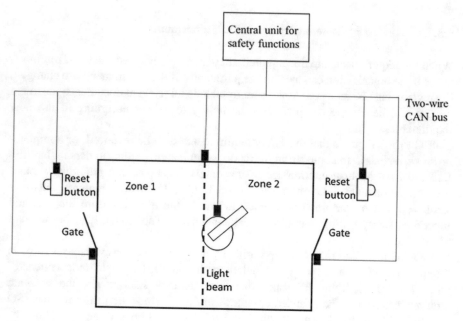

Fig. 5.5 Tool cell diagram

expressed. We need to bridge the gap between the customer/end user and the analyst. The CORE—Controlled Requirements Expression (Mullery 1979)—approach is not mathematically rigorous but aids communication and is designed for requirements expression, not for specification. To quote Mullery: "The central concept of CORE is the viewpoint and the associated representation known as the viewpoint hierarchy. A viewpoint can be a person, role or organisation that has a view about an intended system." See also Table 5.3 (Dick et al. 2004). The example is taken from a tool cell example (Stålhane and Malm 2020) (Fig. 5.5).

ISO 26262 uses customer communication in two ways—between the developing company and subcontractors and between developing company and buyers. When it comes to buyers/end users, it only states that the standard shall provide management of safety anomalies regarding functional safety.

5.3.4   Agile Development: Opportunities and Challenges

Agile methods have several quality assurance mechanisms embedded in the process itself, without any explicit QA role. In principle, the team takes care of quality assurance during sprints and as part of daily stand-ups, sprint reviews, and retrospectives. We have defined SafeScrum, a variant of Scrum with some additional XP techniques that can be used to develop safety-critical software and have the software certified according to the IEC 61508 standard. This imposes a load of additional requirements on the process. In a recent industrial case, we have experienced that the quality assurance mechanisms in Scrum become insufficient. Therefore, we have analyzed the standard, consulted an independent assessor, and worked with the Scrum team to identify necessary additional tasks for a team-internal QA role to be added to the SafeScrum process.

The traditional approach is to place the QA role as a specialized function in the line organization, outside the project. We decided, however, to add the QA role to the Scrum team to be close to the activities and to the information needed to execute quality assurance. This adds to the principle of cross-functional teams in Scrum. We are also considering making this a rotating role to make it a shared responsibility and to share the workload. We need a QA log to trace findings, decisions, corrective actions, and the follow-up/results of these. In our case, Confluence is a good tool to establish this log. We have identified four main tasks:

5.3.4.1   QA Role Task 1: Code Metrics

Check code metric values for all new or changed code: None of the standards provides directions on specific metrics and limits to monitor at the component level. We have consulted the research literature and—as an example—used statistical analysis software to analyze data from code from previous projects in one company to define the following metrics and limits:

1. Number of static paths: 250
2. McCabe's cyclomatic number, $v(G)$: 15. $v(G) = 1 +$ number of predicate nodes
3. Number of parameters: 5
4. Function call count: 13
5. Maximum nesting of control structures: 5
6. Number of executable lines: 70
7. Myer's metric: 10. Myer's metric $= v(G) +$ number of logical operators -1

The collected metrics are displayed, together with their defined maximum values in a radar plot. It is thus easy to if there are metrics that are exceeding their defined limits. If the values are inside their limits, QA will accept the code. If one or more values are outside their limits, the code is presented by QA in the sprint review meeting where the team decides to either accept the violation or plan refactoring. If the violation is accepted, a brief explanation must be added to the log and potentially also in the code as required by the standard. If the violation is unacceptable, the team needs to define a new task in Jira to refactor the code.

ISO 26262-6, Table 9, requires test coverage metrics as part of the test process. EN 50128 refers to a set of metrics that may be used for static code analysis—see Annex A.5 and D.37.

5.3.4.2 QA Role Task 2: Documentation and Code Coverage

Check new/changed code to ensure proper inline documentation and documents. This has to be done manually. In case of missing or poor documentation, the QA log should be updated. If we use an agile approach, the findings should be discussed in the sprint review meeting to decide how to resolve it—giving a task to someone in the team. This check could be done at the end of each sprint

5.3.4.3 QA Role Task 3: Test Coverage (Aka Code Coverage)

The QA log should be updated with references to code not covered by a test. This could be checked by the end of each sprint. Uncovered code should be discussed at the sprint review meeting, and the team should define corrective actions, like defining tasks to produce tests. According to the standard, the test coverage should be at least 99%.

5.3.4.4 QA Role Task 4: Check Requirements-Task-Code Traceability

For new requirements, tasks, and code check that the requirements and code are linked to issues. The QA role should control consistency at the end of each sprint and the team should resolve any identified issues immediately. The IEC61508 standard provides a set of explicit requirements for traceability; see IEC 61508-3, Table A.4: Software design and development—detailed design, and Table A.5: Software design and development—software module testing and integration. We have consulted assessors about a definition of "module" and its size in LOC. Their main point is that "a software module should have a single well-defined task or function to fulfil." An upper limit of 1000 LOC was suggested. If this limit is exceeded, the reason should be explained and documented.

Both ISO 26262 and EN 50128 have strict requirements for traceability. ISO 26262-8 requires that the project maintain traceability between the SRS and the next level—upper and lower. In addition, there shall be traceability between tests and the SRS. EN 50128, annex D.58, requires traceability between requirements, design, and implementation and from these tree artifacts to the tests.

5.3.5 Toward a Quality System

To implement a quality system, the company needs to go through a series of steps. These steps are described shortly below. The three steps are:

1. Development—know your business, list your business activities
2. Implementation—link business activities and quality assurance system
3. Maintenance—improve your quality assurance system based on collected process data using the Plan, Do, Check, Act process cycle

5.3.5.1 Development

In order to develop a good quality assurance system, you first need to understand what your main business activities are and who your customers are. Then you need to list and document your business activities—what you do, why, and how you do it. Once that is in place, you can move on to the implementation phase.

Business activities for a company developing software may include—but are not limited to:

- Customer contacts and requirements acquirement
- Making cost estimates and contracts
- Setting up a project or Scrum team
- Defining /selecting a development process
- Verification and validation
- Hazard log maintenance

5.3.5.2 Implementation

When you have listed all business and administration activities, you can move on to the implementation phase. First, involve the personnel by letting them write down what their jobs cover. Then, collate this in sequences relevant to the list of business activities—from the development phase. Identify the issues where the standard and the business activities list coincide and apply the relevant standards and the quality management system. And remember: keep the quality management system simple, functional, and relevant to the business operations.

5.3.5.3 Maintenance

An important part of QA system management is data collection. The data should provide information from the quality management system, which can be used to improve development and management activities. Important data can be, e.g., the number of customer complaints or the number of errors occurring after delivery.

The data should be analyzed, and the company should decide how to deal with the identified problems. We recommend the "Plan–Do–Check–Act" cycle for improvement. The resulting changes should be monitored so that you know what you have gained, e.g., in efficiency, safety, or reliability.

References

Beck, K., Andres, C.: Extreme Programming Explained: Embrace Change, 2nd edn. Addison-Wesley Professional (2004)

Dick, J., Hull, E., Jackson, K.: Requirements Engineering, 4th edn. Springer (2004)

DNVGL-RP-O101: Recommended Practice. Technical documentation for subsea projects. Ed. June 2016

ERA: Clarification note on safe integration, ERA 1209/063 V 1.0, 2020

Gandy, S.: Why Is Functional Safety An Important Piece of Process Safety Management. May 17, 2017

Hanssen, G.K., Haugset, B., Stålhane, T. Myklebust, T., Kulbrandstad, I.: Quality Assurance in Scrum Applied to Safety Critical Software. XP 2016 Edinburgh

Hanssen, G.K, Stålhane, T., Myklebust, T.: SafeScrum – Agile Development of Safety-Critical Software. Springer December 2018

Infoq: Blog published 2015. www.infoq.com/articles/hardware-can-be-agile Seen 2021-07-16

ISO: ISO 9001 for Small Businesses, 2015

Merritt, R.: What is MLOps. September. 3 (2020) https://blogs.nvidia.com/blog/2020/09/03/what-is-mlops/

MIL-HDBK-2155, MILITARY HANDBOOK: FAILURE REPORTING, ANALYSIS AND CORRECTIVE ACTION, 2014

Mullery, G.P.: CORE – a method for controlled requirement specification. ICSE '79: Proceedings of the 4th international conference on Software engineering September 1979

Myklebust, T.: Certification of Safety Products in Compliance with Directives Using the CoVeR and the CER Methods. ISSC, Boston MA (August 2013)

Myklebust, T., Stålhane, T., Hanssen, G.K., Haugset, B.: Change Impact Analysis as Required by Safety Standards, What to Do? PSAM 12 Hawaii 2014a

Myklebust, T., Stålhane, T., Hanssen, G.K., Wien, T., Haugset, B.: Scrum, Documentation and the IEC 61508–3:2010 Software Standard. PSAM 12 Hawaii (2014b)

Myklebust, T., Stålhane, T., Hanssen, G.K.: Important considerations when applying other models than the Waterfall/V-model when developing software according to IEC 61508 or EN 50128. ISSC 2015 San Diego

Myklebust, T., Stålhane, T.: Safety Stories – a New Concept in Agile Development. SafeComp, Trondheim (2016)

Myklebust, T., Stålhane, T., Bains, R.: The Agile Hazard Log Approach. ESREL (2017)

Myklebust, T., Stålhane, T.: The Agile Safety Case. ISBN 9783319702643. Springer International Publishing (February 2018)

Myklebust, T., Meland, P.H., Stålhane, T., Hanssen, G.K.: The Agile RAMSS Lifecycle for the Future. ESREL, Germany (2019a)

Myklebust, T., Stålhane, T., Hanssen, G.K.: Analysis First Development for Agile Development of Safety Critical Software. ESREL, Germany (2019b)

Papadopoulos, Y., McDermid, J.A., Sasse, R., Heiner, G.: Analysis and Synthesis of the Behaviour of Complex Programmable Electronic Systems in Conditions of Failure. Internationl Journal of Reliability Engineering and System Safety. 71(3), 229–247 (2001)

SCSC-127E Data Safety Guidance version 3.3, DSIWG 2021

Stålhane, T., Myklebust, T.: The role of CM in Agile development of safety-critical software. SafeComp/SASSUR 2015. Delft, Netherlands

Stålhane, T., Hanssen, G.K., Myklebust, T., Haugset, B.: Agile Change Impact Analysis of Safety Critical Software. SafeComp/Sassur (2014)

Stålhane, T., Malm, T.: Four Perspectives on Safety Analysis. ESREL (2020)

Stålhane, T., Myklebust, T.: Agile Safety Analysis. XP 2016 Edinburgh

Wang, Y., Graziotin, D., Kriso, S., Wagner, S.: Communication channels in safety analysis: An industrial exploratory case study, Journal of Systems and Software, April 2018

Wolniak, R., Olkiewicz, M.: The Relations Between Safety Culture and Quality Culture. 10/12/ 2018– CzOTO 2019, volume 1, issue 1

Chapter 6
Safety Analysis Methods Applied to Software

But fortresses were of little value to her afterwards when Cesare Borgia attacked her, and when the people, her enemy, were allied with foreigners. Therefore, it would have been safer for her, both then and before, not to have been hated by the people than to have had the fortresses

—*Nicollo Machiavelli: Il Principe*

What This Chapter Is About
- Early hazard analysis—PHA, HazId, and HazOp
- FMEA and IF-FMEA
- Hazards and sub-hazards
- Risk acceptance—ALARP and GALE
- Dynamic risk and dynamic risk analysis
- Common mode and common cause failures

6.1 Introduction

When analyzing a system, we have three alternatives. We can focus on (1) high or low-level system functionality, (2) system design and architecture, or (3) how the system is built internally. Unless we specify how the safety analysis is to be performed by the manufacturer, there is no way we can be sure of the final result of the analysis. It is also important to be aware that different companies use different descriptions of the same failure condition—e.g., "stuck-at-ON" versus "relay does not open when de-energized." Such differences might be a safety risk when comparing safety analyses from different companies or when different companies analyze different parts of the system. We recommend starting most of the safety analysis projects with a meeting where all participants agree on all system-related and failure-related terms that the different stakeholders will use in the final report.

© The Author(s), under exclusive license to Springer Nature Switzerland AG 2021
T. Myklebust, T. Stålhane, *Functional Safety and Proof of Compliance*,
https://doi.org/10.1007/978-3-030-86152-0_6

6.2   Summary of Safety Analysis Methods

There are a plethora of methods that can be used for safety and hazard analysis. However, the choice and quality of the results will depend on the information available, context, the project, and the participants' experience and knowledge. This section contains a short description and examples of the following methods:

- PHA—preliminary hazard analysis.
- HazID—hazard identification. Mostly used in the early phases of system development.
- HazOp—hazard and operability study. See also IEC 61882:2016 "Hazard and operability studies (HAZOP studies)—Application guide" and SCSC's Data safety Guidance (SCSC 2021).
- FMEA—failure mode effect and analysis. Instead of analyzing a component, we can analyze a function. The table for functional FMEA looks just like the ordinary FMEA but applies the failure modes to a function instead of a component.
- FMEDA—failure mode effect and diagnostics analysis. This has an extension in which it also shows the failure detection probability and the detection methods used.
- IF-FMEA—short for interface-focused FMEA. IF-FMEA shows both possible failures stemming from the component itself and failures due to erroneous input to the component—either from sensors or from other system parts.
- SW-FMEA—software FMEA. Manufacturers can perform SW FMEA at different levels; e.g., for a complex system, it is possible to use data flow diagrams, at a more detailed level (e.g., functions, timing, diagnostics). It is also possible to perform SW FMEA at the code level. Detailed SW FMEA is normally applied after one or more sprints as it has to be based on, e.g., available pseudo code. Possible objectives for Software FMEA include:

  - Identifying missing software requirements
  - Analyzing output variables
  - Analyzing a system's behavior as it responds to a request that originates from outside of that system
  - Identifying (and mitigating) single-point failures that can result in catastrophic failures.
  - Analyzing interfaces in addition to functions
  - Identifying software response to hardware anomalies

- FTA—fault tree analysis is a simple, intuitive method that can be used to analyze safety and reliability—see IEC 61025:2006. We will focus on the safety part.

Several of the aforementioned methods might require estimates of probability and consequence—either graded as "High, Medium or Low" or with a score between 1 and 10. These two scores are used to compute a risk estimate (RPN—risk priority number), which is used to prioritize the identified risks.

6.3   Early Hazard Analysis: PHA, HazId, HazOp, and CHazOp

The PHA is straightforward to use. For each item—HW unit, SW unit, system, or connection, you identify the hazardous conditions and their causes. Then you go on to identify their effects and assess their likelihood, exposure, and magnitude. The main thing with PHA is that it should be done early in the project as it should focus on the interaction between the system and its environment.

The HazId is an alternative, even simpler to use. Here cause and hazardous conditions are crammed into one—the failure condition (Table 6.1).

Risk assessment codes (RAC) are often produced in the matrix through the combination of probability with severity/consequences. The codes reflect risk decisions. The RAC are often presented in five different levels or categories: critical, serious, moderate, minor, and negligible (Table 6.2).

The HazOp method can be used in all phases of system development. The method is heavy on ceremony and requires an experienced process leader. The method applies a set of guide words, also called deviations, to cue in the process participants. Originally, there existed only one set of guidewords but with increased use and practical experiences, the set of permissible guidewords is now quite large. An example of a HazOp is shown in Table 6.3. In the HazOp analysis, each guideword is combined with the selected study nodes—a system element—to initiate a safety consideration. E.g., if we use the guideword "No" and the study node "sensor output," we will get the issue "What happens when there is no sensor output?". The most commonly used HazOp guidewords are "no, more, less, as well as, part of,

Table 6.1   Preliminary hazard analysis (PHA) table

Preliminary hazard analysis Project:_____ Author:_____				Date: ___/___/___ Page ___ of ___	
Item	Hazardous condition	Cause	Effect	Risk assessment Code (RAC)	Assessments
Assigned number sequence	Describe the nature of the condition or include a link to a relevant reference	Describe what is causing the stated condition to exist	If allowed to go uncorrected, what will be the effect or effects of the hazardous condition	Hazard level assignment	Probability of likelihood, exposure, and magnitude

Table 6.2   HazId table

Failure condition	Failure effect	Comments
Trackside equipment down	May lead to two trains on the same track	Signal and trackside equipment should not be on the same system

Table 6.3 HazOp table using standard guidewords

Element: Reset control unit (CU)			Date		
Guide word	Deviation	Possible cause	Consequences	Detected	Actions required
No	Reset not received	Defective reset module or transmission	System stays passive	Not in CU	None
More	More than two status messages are received	Constantly activated switch		No	None
Less	Only one status message is received	Defective CAN (controller area network) controller or reset module	None for normal timing For extreme timing of reset module status messages, the reset could be lost	No	None

other than, and reverse." However, smaller sets such as "no and wrong" are also used in the industry.

Note the "actions required" column. Here we note what should be done in order to handle the identified hazard. This column should include both action and responsible person(s).

Both IEC 61508 and ISO 26262 mention HazOp as an addition or alternative to FMEA. One of the reasons why FMEA is more popular is that it is easier to do. Doing HazOp requires the team to stick to a rather strict process, and many companies feel that it is too much ceremony included.

- IEC 61508-7:2010—The HAZOP technique evolved in the process industry and requires modification for software application. Different derivative methods, e.g., Computer HazOp—"CHazOp," have been proposed which in general introduce new guide words and/or suggest schemes for systematically covering the system and software architecture.
- ISO 26262-3:2016—Each functional safety requirement shall be specified by considering the following... This activity can be supported by safety analyses (e.g., FMEA, FTA, HazOp) in order to develop a complete set of effective functional safety requirements.

According to IEC 61508-7:2010: "A team of engineers, with expertise covering the whole system under consideration, participate in a structured examination of a design, through a series of scheduled meetings. The team considers both the functional aspects of the design and how the system would operate in practice (including human activity and maintenance)."

The information we present for CHazOp is taken from Raman (2005). CHazOp may be viewed as an extension of HazOp to root cause level in that in HazOp we stop with the deviation being a control loop failure (high, low, or none), while in CHazOp, we extend this failure to its causes in the PES. The main information required for CHazOp is:

Table 6.4 CHazOp guidewords

Deviation Guideword	Interacting subsystem
	Communication
No	Signal (zero read, full-scale read)
More	More current. Erratic signal
Part of	Incomplete signal
Other than	Excessive noise. Corrupt signal
Early	Signal generated too early (timer problems)
Late	Signal generated too late
Before/after	Incorrect signal sequence
	Digital hardware
No	I/O failure
More	Multiple failure (control card, processor rack, processor)
Part of	Partial failure of card, failure of counters
Other than	Abnormal temperature, dust
	Software
No	Program corruption
More	Memory overflow
Part of	Addressing errors/data failure
Other than	Endless loops, data validation problems, operator override
Early/late	Timeout failure, sequence control problems, sequence interpretation error

- SIS loop diagrams or block diagrams or flow charts
- Electrical circuit diagrams where relevant
- Instrument cause and effect charts and
- P&I diagrams for identifying the process consequences of deviations in PES

The CHazOp focuses on the interacting subsystems, and the guidewords used are as follows (Table 6.4):

6.4   Generic, Domain-Specific Hazard Lists

An alternative to identifying a set of hazards from a HazOp is to use a set of generic hazards. The set of generic hazards might be domain specific. If we decide to use generic hazards, we need to relate them to the system that we want to analyze. We can also use the concept of sub-hazards to make a more precise hazard analysis as shown in Fig. 6.1.

We need a list of relevant generic hazards and a description of the system to be analyzed—its functionality and the environment where it will be operating. There are three sources for generic hazards—literature, standards, and the company's own hazard log. For example, it is possible to, e.g. start with a generic hazard list and add

Fig. 6.1 From generic to system-specific hazards

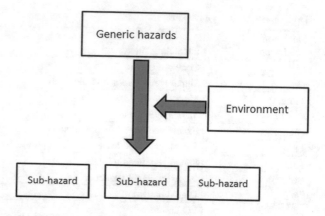

the company's own experiences as they become available. Proof of compliance for generic risk assessment should be part of the total risk assessment for the system under consideration.

If we are dealing with a system containing several hazard sources—e.g., hazardous materials, buildings, environment, and software, there are two ways to attack this problem. First, if it is not already available, we could construct a generic fault tree for the total system, just as is done for a building in the diagram in Annex A. We can then assign a relevant generic hazard list to each part of the fault tree. This process will give a good overview of the relevant hazards. A simple alternative is to break the system down into its individual components and apply the relevant hazard list to each.

For software hazards not related to the application domain, language, or method, M. Squair has published an important paper (Squair 2014) containing a generic list of software hazards, together with a list of error-prone coding semantics and software countermeasures.

According to IEC 61508-1, the hazard and risk analysis shall consider the following:

- Each determined hazardous event and the components that contribute to it. This is where we will use the generic hazards and the hazard log.
- The consequences and likelihood of the event sequences with which each hazardous event is associated. This must be based on system and operational experience and knowledge.
- The tolerable risk for each hazardous event.
- The measures are taken to reduce or remove hazards and risks—e.g., design and the inclusion of barriers.
- The assumptions made during the analysis of the risks, including the estimated demand rates and equipment failure rates; any credit taken for operational constraints or human intervention shall be detailed.

ISO 26262-2 describes the hazard analysis as follows: The goal is to judge whether the results of the hazard analysis and risk assessment and the methods used are convincing and are supported by rationales, as well as to judge whether the safety goals cover all identified hazardous events that are classified with an ASIL.

The main part of the document is the system description and the hazard analysis done based on the generic hazards. In order to show proof of compliance, we need a reference to the relevant list of hazards, why this hazard list was chosen, which other lists were considered, identification of the personnel who participate when it was done, and the process used.

In addition, we need to document how we have handled the identified hazards—e.g., by introducing barriers or changing the design or architecture. The PoC of this is found in the relevant documents—e.g., in the software architecture and design verification report.

6.4.1 Output

The main output from this activity is the filled-in hazard descriptions and descriptions of how they have been handled. In addition, it is practical to report which part of the generic hazard lists was used to identify which hazard. This information should later be used to update the hazard log and to improve the hazard identification process.

6.5 FMEA and FMEDA

The FMEA method has many things in common with HazOp but is not so heavy on ceremony. The purpose of an FMEA will depend on when in the development process the analysis is performed. Early in the process, we might use FMEA on the design. Later on, we might use FMEA on the system structure. Both of these early FMEAs might give input to the system safety requirements. FMEA uses failure modes instead of guidewords to help the participants to focus. We need to remember that a failure mode is not an error but a way a component can fail. As long as we keep that in mind, any set of failure modes might do. We recommend the use of a generic set of failure modes. Table 6.5 shows the set of generic failure modes recommended by the Nuclear Regulatory Commission (NRC).

A simple alternative that can apply to all types of components has been used by, e.g., Airbus (Tables 6.6, 6.7 and 6.8):

- Loss—no output, no action
- Erroneous—wrong output, no action

FIT (failure in time) is the sum of safe and dangerous failures (1×10^{-9} failures per hour).

Table 6.5 NRC set of generic failure modes

ID	Failure mode	Elaboration	Remarks
A1	Fail to perform the function at the required time	Deviation from requirement in time domain	Omission, No action, No output, Reacts too late
A2	Fail to perform the function with correct value	Deviation from requirement in value domain	Wrong output
A3	Performance of an unwanted function	Deviation from expected performance	Commission, Wrong action
A4	Interference or unexpected coupling with another module	Deviation from expected system performance due to module interaction	Commission

Table 6.6 FMEA table

Unit description		Failure description			Failure effect	Recommendations
Function	Operational conditions	Failure mode	Failure cause	Failure detection	On the next level	

Table 6.7 IF-FMEA table

Unit description	Failure description			Failure effect	Recommendations
	Failure mode	Component failure cause	Input failure cause		
Controller	No action	Power loss	No signal from sensor	No heat	Feedback switch info to controller for verification

Table 6.8 FMEDA table

Component information	Failure modes	Effect	FIT	Failures			Diagnostic method
				Safe	Dangerous	Detected	
Total number of FIT							
Failure rates							

6.6   Fault Tree Analysis

The fault tree analysis (FTA)—is based on a simple idea. We start by asking "how can this system fail?" The answers will be documented in a tree fashion using "AND" and "OR" gates to show how events may be combined. We can then ask the same question for each of the components on the next level and so on. The result is an easy-to-interpret diagram as shown in Fig. 6.2. When used in safety analysis we need to answer the following question for each failure: "How can we prevent this failure?" The answer will usually give rise to a barrier or extra testing of the

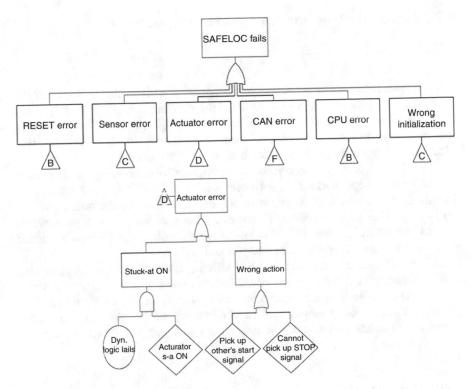

Fig. 6.2 Fault tree analysis example

component involved. The fault tree will also show what the most pressing problem is.

Figure 6.2 shows that the actuator will fail if we either get a wrong action or the actuator is stuck at ON (Stålhane and Malm 2020). However, a wrong action can be caused by one out of two independent failures, while the stuck-at ON only can happen if we get both the actuator stuck-at-ON and the dynamic logic fail

It is possible to do a simple assessment of failure probability from a fault tree using a qualitative reliability assessment (high/low) and the following rules:

1. Assign a reliability score of "+"—high—and "−"—low to each bottom event.
2. For an OR-gate: all inputs are low => output is low. For all other cases, the output is high.
3. For an AND-gate: one input is low => output is low. For all other cases, the output is high.

6.7 Common Mode and Common Cause Failures

6.7.1 Introduction

Independence between functions, subsystems, or items is often required to satisfy the safety requirements. It is also important for the development of resilient systems. Common cause analysis (CCA) helps us to verify independence or to identify specific dependencies. The CCA also supplements the FTA—see IEC 61025:2006—when the FTA is used to assess the probability of a failure event since it might be necessary to consider couplings between seemingly independent events. FTA is recommended for software both in IEC 61508-3:2010 and in ISO 26262-6:2018. Common mode and common cause analysis are important both for hardware and for software. We will, however, focus on the use of these methods applied to software. As you will see below, this has ramifications for how we define common mode and common cause analysis and failures.

IEC 61508-3 discusses common cause failures but only if the test system uses diverse software or if the common cause failure can be initiated from an external event. ISO 26262-6 mentions common cause failures in Annex E, where they discuss software failures propagating from one component to another. Otherwise, common mode and common cause failures for software are treated in a rather superficial way.

6.7.2 Definitions

Common mode and common cause failures are subsets of dependent failures. Their relationship is as shown in Fig. 6.3 taken from Rausand and Høyland (2004).

For common cause failures and common mode failures, we will use the following definitions—see (Malm and Hietikko 2000):

- **Common Cause Analysis (CCA)**: used to identify an event or a failure, which bypasses or invalidates redundancy or independence. A common cause failure (CCF) is a failure of different items, resulting from a single event, where these failures are not consequences of each other.
- **Common Mode analysis (CMA)**: used to identify an event which affects a number of elements otherwise considered independent. Common mode failures (CMF) are failures of items characterized by the same fault mode.

Note that some standards—e.g., IEC 61508-4:2010 and IEC 61511-1: 2016—also have definitions of common cause failures. However, these definitions are too narrow as they only include CCFs due to failures in separate channels or devices. See, for instance, the IEC 61508 definition—cursive added by authors: "3.6.10 common cause failure—failure that is the result of one or more events, causing concurrent failures of *two or more separate channels* in a multiple channel system,

Dependent	Failure		The probability of a group of events which probabilities cannot be expressed as a simple product of unconditional probability of failure of single components.
	Common Cause Failure		This is a kind of dependent failure which occurs in redundant components in which a single common cause - simultaneously or near simultaneously leads to failures in different channels.
		Common Model Failure	This definition applies to failures of common causes in which multiple elements fail similarly in the same mode.
	Cascade Failure		These are all dependent failures that do not share a common cause, meaning they do not affect redundant components.

Additionally:
The definition of dependent failures" includes all definitions of failures that are not independent. This definition of dependent failures clearly implies that an independent failure in a group of events can be expressed as a simple product of conditional probabilities of failures of a single event.

Fig. 6.3 Dependent failures

leading to system failure." IEC 61508-6, Annex D has a detailed approach for handling CCF but only for hardware.

6.7.3   Common Mode Analysis (CMA)

The CMA is a qualitative method used to ensure the strength of the design and contributes to the verification that independence principles have been applied when necessary. Independence can be checked by developing a fault tree and then verifying that ANDed events in the fault tree—events that are input to the same AND gate—are independent in the implementation. This should include failures in system components that jeopardize their independence assumption. The situation can be described by Fig. 6.4. The fault tree on the left-hand side describes the situation if the two events A and B are independent, while the fault tree on the right-hand side describes the situation if there also can be a common cause failure.

For the case on the left-hand side, we have that

$$P_{TOP} = P_A * P_B$$

Fig. 6.4 The influence of a CCF (common cause failure)

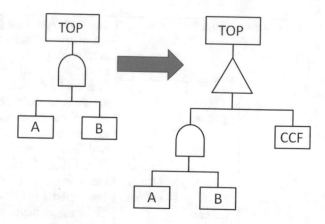

Table 6.9 Typical β-values

Application	Beta values
Electrical equipment	$\beta = 0.01$ (best) 0.30 (worst)
NPP data	$\beta = 0.03$–0.22 with 0.10 (average)
Safety systems	$\beta = 0.001$–0.05 (good engineering) to 0.25 (poor engineering)
Hardware failures	$\beta = 0.001$–0.10
General applications	$\beta = 0.01$–0.1 (good engineering) 0.25 (poor engineering)

If we assume that the two components A and B are equal, consider common cause failures and use the β-factional model, we can write

$$P_{TOP} = P_C(1 - \beta) \times P_C(1 - \beta) + P_C\beta$$

If we have P_A and P_B equal to 10^{-3} and $\beta = 0.10$, we get

$$P_{TOP} = 10^{-3} \times 0.90 \times 10^{-3} \times 0.90 + 10^{-3} \times 0.10 = 0.81 \times 10^{-6} + 10^{-4}$$

$$P_{TOP} = (1 + 0.0081) \times 10^{-4}$$

Since there is no commonly agreed way to assign failure probabilities to a software component, it is challenging to use this approach on a software system. One possible way out of this could be to use the failure probability related to the component's SIL-value, but there is no reported experience on this approach.

Bao et al. (2020) have collected typical β-values from the literature. Their results are shown in Table 6.9. One challenge when using these data is how to differentiate between good and poor engineering. Note: NPP data are data from a nuclear power plant.

Another way to assess the β-values can be found in IEC 62061:2021, Annex E (Table 6.10).

Table 6.10 Checklist for assessment of the β-factor, © IEC

Separation/segregation	Score
Are SCS (Safety-related Control Systems) signal cables for the individual channels routed separately from other channels at all positions or sufficiently shielded?	5
Where information encoding/decoding is used, is it sufficient for the detection of signal transmission errors?	10
Are SCS signal and electrical energy power cables separate at all positions or sufficiently shielded?	5
If subsystem elements can contribute to a CCF, are they provided as physically separate devices in their local enclosures?	5
Diversity/redundancy	
Does the subsystem employ different electrical technologies, for example, one electronic or programmable electronic and the other an electromechanical relay?	8
Does the subsystem employ elements that use different physical principles (e.g., sensing elements at a guard door that use mechanical and magnetic sensing techniques)?	10
Does the subsystem employ elements with temporal differences in functional operation and/or failure modes?	10
Do the subsystem elements have a diagnostic test interval of ≤ 1 min?	10
Complexity/design/application	
Is cross-connection between channels of the subsystem prevented with the exception of that used for diagnostic testing purposes?	2
Assessment/analysis	
Have the results of the failure modes and effects analysis been examined to establish sources of common cause failure and have predetermined sources of common cause failure been eliminated by design?	9
Are field failures analyzed with feedback into the design?	9
Competence/training	
Do subsystem designers understand the causes and consequences of common cause failures?	4
Environmental control	
Are the subsystem elements likely to operate always within the range of temperature, humidity, corrosion, dust, vibration, etc. over which it has been tested, without the use of external environmental control?	9
Is the subsystem immune to adverse influences from electromagnetic interference up to and including the limits specified in Annex E?	9

Table 6.11 β-values based on the checklist above	Overall score	Common cause failure factor (β)
	< 35	0.10
	35–65	0.05
	66–85	0.02
	86–100	0.01

The checklist is used as follows: Answer the questions in the checklist, add the scores for all Yes-answers and decide on the ß-factor using Table 6.11.

CMA can also be used to derive safety requirements. This requires that we first develop a fault tree for the relevant parts of the system. The CMA, as done in the SafeLoc project (see Stålhane and Malm 2020), is performed in the following way:

1. Identify the CMA requirements. This activity includes the identification of possible couplings and corresponding barriers to common mode failures. The identification is done by including all events that are input to an AND gate. It will be useful to consider the four categories of AND-gates shown in Klim and Balazinsky (2007)

    a. Category 1 AND gate: redundant channels and/or including Command/Monitor of the architecture of the electronic units.
    b. Category 2 AND gate: a model for the combination of a mechanical hardware failure and obviously independent electronic failure.
    c. Category 3 AND gate: a model for the redundant hardware design.
    d. Category 4 AND gate: a model to address a sequence of events (e.g., item #1 fails first, item #2 fails second) or to build the general fault tree architecture.

2. Analyze the system architecture and components to check that all requirements identified in activity 1 are fulfilled.
3. Document the results—a list of common modes.

There has been some work on getting rid of CMF in software systems using N-version programming combined with a voter (see, for instance, Thomson 2012). However, experiments have shown that this is not an efficient approach to get rid of CMF. To quote (Brilliant et al. 1990):

> Correlated failures occur when the partial functions computed by the paths are identically wrong. The actual mistakes made, however, need not be similar or logically-related. We did find that programmers often make identical errors in logic. Any given algorithm for solving a problem is likely to involve some computations that are simply more difficult to handle correctly than others, and programmers are more likely to make mistakes on difficult computations than easy ones. We also found, however, that correlated failures arise from logically-unrelated faults in different algorithms or in different parts of the same algorithm.

6.7.4 Common Cause Analysis

Safety instrumented systems (SIS) often include redundancy to enhance reliability and safety, but the intended effect may be reduced when common cause failures are taken into account.

Brancati (2019) suggests the following approach for identifying common cause failures:

- Step 1. A set of guidewords for events or root causes that may be a cause of common failures of software elements is identified (see Table 6.2).
- Step 2. Identify couples of SW components (see coupling factors in Table 6.2). Typically, these are elements that:

Table 6.12 Examples of defenses, root causes, and coupling factors

Defenses	Root causes	Coupling factors
• Redundancy/diversity • Separation • Understanding • Analysis • Operator interaction • Safety culture • Environmental control • Environmental testing • Operational testing (OEDR)	• State of other components • Design, manufacturer, or construction adequacy • Design related reviews inadequacy • Design and functional measures • Internal to component, piece part • Maintenance • Procedure inadequacy • Human actions • Training, expertise, and competence • Other unknown	• Environmental control • Environmental external • Environmental internal • Hardware • Hardware design • Software quality deficiency • Operational • Operational procedure • Operations staff • Operational follow-up and analysis • Modifications, upgrades • Maintenance/test (M/T) • M/T staff and procedure • Physical and layout measures

Realize safety-critical functionalities through software diversity.

Are replicated software, running on the same hardware.

Implement redundant functionalities.

This item includes the redundancy of a safety mechanism with respect to a target element.

- Step 3. A guidewords-based analysis is applied to each set, to understand the impact of such failures from a system-level point of view.

A set of root causes, coupling factors, and defenses made from a combination of information from (Lindberg 2007) and (Hauge et al. 2016) are shown in Table 6.12. It might be necessary to construct new terms for common cause analysis of software systems.

From Figure, we see that the most important factor for CCA is the coupling factor. This is the most important item to identify and the most important item to deal with in order to avoid CCF (Fig. 6.5).

The general approach for analyzing common cause failures and the defense can be as shown in Fig. 6.6. Note that coupling factors, defense, and root causes will depend on factors such as type of system, area of operation, type of environment, and possible operator influence. Several reports have published data on common coupling factors and root causes. An example is the ICDE project report (see ICDE 2008). This report states that the most important coupling factors are:

- Maintenance/test schedule: 43%
- System design: 15%
- Hardware: 10%
- Maintenance/test procedure: 7%

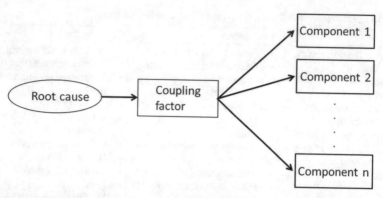

Fig. 6.5 The role of the coupling factor in CCA

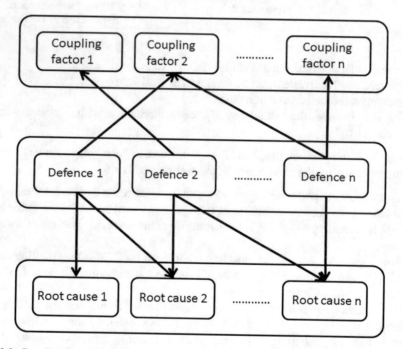

Fig. 6.6 Coupling factor, defense, and root causes. © (Lindberg 2007)

These four factors make up 30% of all factors and account for 75% of the effect. Thus, Pareto's rule holds. The report also gives data for root causes. However, this information is difficult to use since 44% of all root causes are categorized as "Other."

Figures 6.7, based on Lindberg (2007), shows an example of a coupling factor, defense, and root cause diagram.

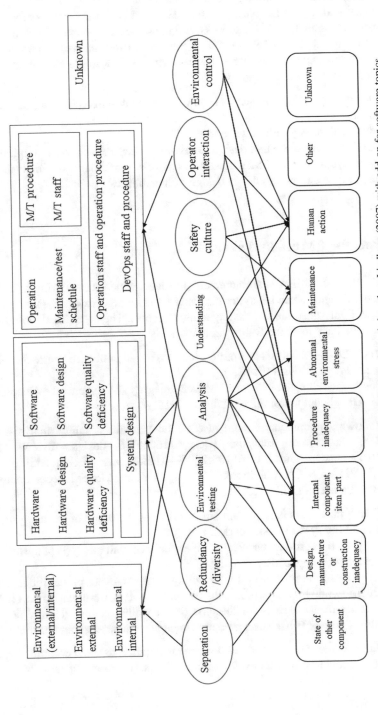

Fig. 6.7 Coupling factor, defense, and root cause diagram—example. The figure is based on Lindberg (2007) with add-on for software topics

According to Leedeo (2020), the most important coupling factors stem from using the same design principles, the same HW and SW, the same operation and maintenance personnel, the same process, and the same environment. Therefore, you should also consult IEC 61508-3: 2010, Annex F (informative) Techniques for achieving noninterference between software elements on a single computer.

The coupling may be reduced by:

- Introducing separation and segregation of redundant items (physical, functional, electrical)
- Introducing diversity in hardware and software
- Simplifying architecture and design, to avoid having undiscovered couplings.
- Using analyses such as FMECA, zonal analysis, particular risks analysis, and common mode analysis to detect design vulnerabilities

6.8   PoC for FMEA and IF-FMEA

First and foremost, FMEA is an incremental process which starts during high-level design and only ends when the system is ready for shipping. However, for the safety case, we will only need the final safety analysis report. Therefore, as standards and guidelines, we have used the following documents:

- EN 60812:2018—Analysis techniques for system reliability procedure for failure mode and effect analysis (FMEA)
- Ontology Modelling in Physical Asset Integrity Management—Chapter 3: FMEA, HazId, and Ontologies (Stålhane 2015)

The following documents should also be relevant for a safety analysis: IEC 61025:2006—fault tree analysis, IEC 61160:2006—design reviews, IEC 61882:2016—HazOp, and IEC 61165:2006—Markov analysis. In addition, the "AIAG & VDA FMEA Handbook: Potential Failure Mode Effects Analysis, FMEA" will be useful (AIAG 2019).

- The FMEA plan. This is important since it defines the scope of the analysis, the system, and the system's boundaries.
- The analysts' background, especially the experience with this type of system and with the relevant application domain(s).
- A description of the rules used for assigning values to event probabilities, event consequences, and how to combine the two values—e.g., a criticality matrix. This also includes decision rules for the type of controlling actions that are needed. IEC 60812, Annex B has good guidelines and several approaches for this.

In addition to the abovementioned documents required to show compliance, it is important to decide on a strategy for failure modes. There are at least two main approaches:

- Define failure modes for each component participating in the analysis, based on what each component does
- Use a set of standardized failure modes

There are several types of FMEA—e.g., IF-FMEA (interface-focused FMEA), FMECA (failure mode effect and criticality analysis), FMEDA (failure mode effect and diagnostic analysis), and DFMEA (design FMEA). When assessing the quality of an FMEA, the following information is important:

References to the Input Documents

- The failure modes used
- The terms used and showing that they are used consistently
- The components—hardware and software—considered
- A description of the actions done to handle the dangers identified by the FMEA, including a reference to the test report

The output documents fulfill two roles: to show what has been done to assure system safety, and to show that this has been done in a way that demands confidence in the results. This process has the following outputs: a document that shows that:

- The FMEA has been performed by the manufacturer according to all the requirements in IEC 60812, part 5.3—"Perform the FMEA."
- The whole process has been under quality control—see also the information specified in the previous section.

An FMEA is documented by the manufacturer in an FMEA table. The layout of the table will depend on what is considered in the analysis. However, the table shown in Table 6.13 is fairly common. The recommendation field in the FMEA is important since it is used to register three types of actions, listed in order of priority:

- Prevention. Change the design or implementation to remove or reduce the probability of the failure's occurrence => preventing a risk (potential problem) from becoming a real problem.
- Handling. Prevent the failure's consequences by checking output, using time out, or other control activities.
- Reduction. Reduce or control the failure's consequences.

The barriers are identified by looking at how a failure propagates through the system in order to create a hazardous event (Fig. 6.8).

Note that the barriers do not have to be software related. They can also be physical barriers, e.g., a fence, operational barriers, or implemented as part of a user manual.

The idea of the interface-focused FMEA (IF-FMEA) stems from the HIP-HOPS project (Papadopoulos and McDermid 1999). Some things are, however, different (Table 6.13):

- The output from the process is a failure mode. This enables us to couple the result of this FMEA to an FMEA of another component.

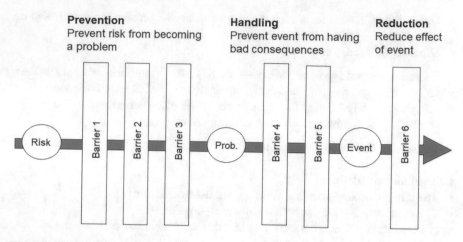

Fig. 6.8 Barrier model

Table 6.13 Interface-focused FMEA

Component ID		List of input sources	
		• Input 1	
		• Input 2	
		• ...	
Output failure mode	*Description*	*Input deviation*	*Component malfunction*
FM1	Description of FM1	Input deviations that can cause FM1	Component failures that can cause FM1
FM2	...	...	...

- The component can fail in two ways: something is wrong with the component, or it gets wrong/bad input.

6.9   Hazards and Sub-hazards

Before we look at the hazard analysis methods, we need to look at why hazards can be decomposed and added. Generally, competing risks are a set of risks where one of them will eventually happen—basically, they are competing for the user's attention. The term is often used for a risk that either

- Hinders the observation of the event of interest or
- Modifies the chance that this event occurs

Example: when studying the life of a roller-baring in a machine, an axel that breaks is an event that competes with the event of interest—the roller-bare failure.

Consider a situation where we have some equipment where several things can go wrong and create a hazard. Once the equipment is started, it can fail in one of k different ways, described as $\{m_1, m_2, \ldots, m_k\}$. We will assume that the failure times are independent. Thus, failure j will occur at time x_j, and the time to an equipment failure is given as $\min\{x_1, x_2, \ldots, x_k\}$. Note that we in each case only will only observe the first failure, not the other ones. That is why it is called competing risks.

We will denote the cumulative distribution of x_i by $F_i(x)$. Thus, for the first failure we get:

$$F_x(x) = 1 - \prod_{i=1}^{i=k} [1 - F_i(x)]$$

If we denote the hazard corresponding to failure mode m_k as $h_k(t)$, we can write

$$F_x(x) = 1 - e^{-\int_0^x h(S)ds} \quad \text{and thus} \quad F_x(x) = 1 - e^{\int_0^x \sum_{i=0}^k h_i(S)ds}$$

From this, we see that $h(S) = h_1(S) + h_2(S) + \ldots + h_k(S)$. In other words, the system's hazard is equal to the sum of the hazards from each failure mode—the sub-hazards. Furthermore, we have that

$$1 - F_x(x) = e^{-hx}$$

The equation above is the standard formula for an exponential distribution. This shows that under the conditions stated above, all failure times are exponential even when we have several possible modes of failure. The competing failure model can only be used if all the failure modes are independent. If we, in addition, assume that each hazard rate is constant, we get the following simple expression:

$$h - h_1 + h_2 + \ldots + h_k$$

This result is also a basis for the part-count model (see MIL-HDBK-217F 2018). The idea is simple. If the probability of failure for component i in the equipment is P_i and the components fail independently, then the probability of equipment failure is $P = \Sigma P_i$. If the components do not fail independently, then the part-count model gives the upper limit for equipment failure.

The idea of competing risks and the result of $h = h_1 + h_2 + \ldots + h_k$ allow us to decompose a hazard into sub-hazards. Using hazards and sub-hazards allows us to use the idea of stepwise refinement in hazard analysis. The math presented at the start of this chapter shows that we are on safe ground as long as we can apply the concept of competing risk.

An example of how to present an analysis using sub-hazards in Table 6.14. It is part of the hard analysis of a wind turbine and is based on the work of Puisa (2019). The first step is to formulate accidents and describe how these accidents can occur. Note that accidents correspond to undesirable deviations from the system objective

Table 6.14 Part of the hazard analysis of a wind turbine

Mode of operation	System-level hazards and ID
:	:
:	:
Rest	H5: Position and/or heading are not maintained (drive-off, drift-off) within the predefined ranges before an operation is completed. H6: Station keeping capability does not match the operational requirements of the vessel.
Interface with turbine	H5, H6
Interface with daughter craft	H5, H6

while hazards correspond to violated actions that are needed to achieve the objective (Table 6.14).

The sub-hazards that make up H5 are listed as follows:

- H5.1: Set points are not achieved in required time.
- H5.2: Set points are not maintained within alarm limit.
- H5.3: Communication between thruster and remote controller is not maintained at required frequency.
- H5.4: Loading of el. motors and/or diesel engines exceeds the limits.

For the railway domain, relevant system-level hazard could be train collision, derailment, and fire. Sub-hazards for, e.g., derailment would be broken rail, that could cause derailment.

6.10   Risk Acceptance: GALE and ALARP

First two definitions:

- GALE (Globally At Least Equivalent)—we have done enough to know that the new risk level is not higher than before the change. In order to use GALE, we need to assess our current risk. This approach will help us to understand our current situation.
- ALARP (As Low As Reasonable Practical)—we have done all that is reasonable to prevent dangerous events. A reasonable goal is to reduce risk until the extra mitigation cost exceeds the risk reduction value achieved. This leads to a strategy where we assess all risks, rank them, and mitigate as many of the risk causes as economically reasonable.

We will start by looking at GALE (see Halbert and Tucker 2006).

Before doing a GALE assessment, we need to consider the following rules:

Table 6.15 Frequency index

Frequency classification	Occurrences per year		F_E
Very frequent	200	Every day	6
Frequent	100	Every few days	5
Probable	40	Every week	4
Occasional	10	Monthly	3
Remote	1	Annually	2
Improbable	0.2	Every few years	1
Incredible	0.01	Once in 10 years	0

Table 6.16 Probability index

Classification	Interpretation	P_E
Probable	It is probable that this event, if it occurs, will cause a problem	3
Occasional	The event, if it occurs, will occasionally cause a problem	2
Remote	There is a remote chance that this event, if it occurs, will cause a problem	1
Improbable	It is improbable that this event, if it occurs, will cause a problem	0

- Every important hazard must have been identified.
- The overlap between the identified hazards must be as small as possible.
- Since GALE is all about looking at the effect of a considered change, the hazards must be split into three disjoint sets:

  Hazards that apply to both the status quo situation and the situation after the change.
  Hazards that are unique for the situation after the change.
  Hazards that are unique for the situation before the change.

We can define the risk as $R = C \times P$ (accident | hazard) $\times P$(hazard).

From this expression, we see that we can reduce the risk in two ways: by focusing on the hazard—prevention, or by preventing the hazard from causing an accident—control. We introduce the following symbols:

- H_f: hazard frequency. $\log(H_f) = F_E$
- p_H: the probability that the hazard will lead to an accident. $\mathrm{Log}(p_H) = P_E$
- f: the accident frequency. $f = H_f \times p_H$.
- C: accident severity. $\log(C) = S$
- R: risk. $R = C \times f$ or $R = C \times H_f \times p_H$.
- $I_E = \log(C) + \log(H_f) + \log(p_H) + c$
- $I_E = F_E + P_E + S$

The constant c used in the risk index I_E is needed in the general case because the risk measure is relative. In GALE analysis, it is common practice to use tables where probabilities, frequencies, and consequences each are separated by a decade. The following three tables show examples of index tables for the three indices. We will denote the risk index for hazard j as I_j (Tables 6.15, 6.16 and 6.17).

Table 6.17 Severity index

Severity classification	Interpretation	S
Severe	The portion of occurring problems that have serious consequences is much larger than average	2
Average	The portion of occurring problems that have serious consequences is similar to our average	1
Minor	The portion of occurring problems that have serious consequences is much lower than average	0

The global risk index (I_{GR}) can be found by summing the total risk of all identified risks. The risk is $10^{\text{risk index}}$. Thus, we have that $I_{GR} = \log \left\{ \sum 10^{I_{Ej}} \right\}$. The global risk index is a measure of the total risk of the system.

We can use the GALE concept in two ways:

1. To rank the hazards. We should try to reduce the hazard with the largest I_E first. In addition, the text in each table tells us what we have to aim for if we want to reach a certain I_E-value.
2. We can use the I_{GR} as a hazard requirement. It is then up to the developer company to identify all hazards and then show that the total index (I_{GR}) is below the required value.

What follows is a short description of ALARP.

ALARP stands for "As Low As Reasonably Practicable." The term "Reasonably Practicable" has a legal definition, at least in the UK (see Ross 2019).

1. "Reasonably practicable" is a narrower term than "physically possible."
2. A computation must be made by **the owner**, in which the quantum of risk is placed on one scale and the sacrifice involved in the measures necessary for averting the risk is placed in the other.
3. If there is a gross disproportion between them—the risk being insignificant in relation to the sacrifice—the defendants discharge the onus on them.
4. This computation shall be made by **the owner** at a point of time anterior to the accident.

The term "Gross disproportion," used in issue 3 in the definition above, has an agreed-upon definition, as follows:

Gross disproportion—Definition

If a measure is practicable and it cannot be shown that the cost of the measure is grossly disproportionate to the benefit gained, then the measure is considered reasonably practicable and should be implemented. Thus, the criterion is reasonably practicable, **not reasonably affordable**: justifiable cost and effort are not determined by the budget constraints/viability of a project.

ALARP uses Table 6.18 to map event probability and event consequences into three areas: intolerable, tolerable, and broadly acceptable.

Once the area of tolerability is decided, Fig. 6.9 shows how we shall handle the risks—IEC 61508-5:2010.

Table 6.18 ALARP consequence, probability, and tolerance table

Probability	Consequences				
	Single fatality	2–10 fatalities	11–50 fatalities	50–100 fatalities	More than 100 fatalities
Likely >10⁻²	Intolerable	Intolerable	Intolerable	Intolerable	Intolerable
Unlikely 10⁻⁴–10⁻²	Tolerable	Tolerable	Intolerable	Intolerable	Intolerable
Very unlikely 10⁻⁶–10⁻⁴	Tolerable	Tolerable	Tolerable	Tolerable	Intolerable
Remote <10⁻⁶	Broadly acceptable	Broadly acceptable	Tolerable	Tolerable	Tolerable

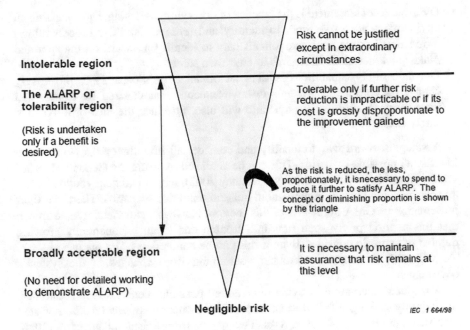

Fig. 6.9 ALARP tolerability and activity diagram. © IEC

For risk reduction measures, we should apply the principles of prevention as a hierarchy:

- Elimination of risk by removing the hazard
- Substitution of a hazard with a less hazardous one
- Prevention of potential events
- Separate people from the consequences of potential events
- Control of the magnitude and frequency of an event

- Mitigation of the impact of an event on people
- Emergency response and contingency planning
- Resilience planning

6.11 Dynamic Risk and Safety Analysis

6.11.1 Introduction

We will start by defining the two closely related concepts: dynamic risk and dynamic safety.

- Dynamic risk assessment is the practice of observing, assessing, and analyzing an environment while we work to identify and remove risk. The process allows individuals, robots, and autonomous cars to identify a hazard on the spot and make quick decisions in regards to their own safety.
- Dynamic safety capability describes an organization's capacity to proactively change its core safety systems in environments characterized by change and uncertainty. Making safety dynamic will also introduce the idea of a dynamic safety case.

A straightforward way to identify and consider all important risks would be to identify all possible scenarios. This can be achieved by using the General Morphological Analysis (GMA) method (see Ritchey 2002), which will help you to identify all scenarios and thus all important but hitherto unknown scenarios. There is a short presentation of GMA at the end of this section. However, experiences have shown that this method pretty soon runs into trouble due to an exponentially growing number of states to consider. Even if this can be handled, e.g., by the use of tools, we will still be facing the problems stemming from changes in the system's environment.

A practical alternative to trying to create all possible scenarios is to create new scenarios by looking at the information provided by near misses and risks associated with other, previously observed risks. Two quotes from Kristamuljana et al. (2018) are important here:

- One of the main things learned in risk management since the new millennium is the previously unobserved levels of correlation (between old and new risks and between old and new accidents)
- Traditional risk models fail to recognize the interconnections between risks or the effect of clustering risks. A seemingly low risk could potentially form part of a high severity risk cluster, that of operational risks.

If we want to identify hitherto unknown but dangerous scenarios, there are two methods to consider—preferably used together:

- DyPASI: Dynamic Procedure of Atypical Scenarios Identification is a self-learning method based on a systematic review of past accidents and near misses.
- ERMF: Emerging Risk Management Framework or Enterprise Risk Management Framework.

6.11.2   Dynamic Safety and Risk

The idea of dynamic safety and risk analysis is pretty obvious, at least at the surface of things. You do a risk assessment or a safety analysis and when the environment changes, you need to repeat the assessment or analysis. This is the case both during development—handled through an agile development approach—and later during operation using DevOps. In order to make this work, it is important to have a process in place that will help us move from experience and observations to safety performance. Figure 6.10 is adapted from Griffin et al. (2016) and shows the dynamic safety capability and how we move from experience to dynamic safety capability and finally to safety performance.

The terms used should be understood as follows (Zollo and Winter 2002):

- Experience accumulation describes the accumulation of experience through tacit learning from ongoing actions and events.
- Articulation describes the process through which implicit information from events is articulated and shared through collective discussions and processes of sense-making.
- Codification describes the understanding that is formalized in tools such as manuals, blueprints, and software and in formal organizational and regulatory procedures.

An alternative is the four-step model suggested by Coachio (2019), which runs as follows:

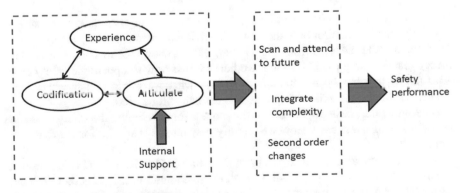

Fig. 6.10 Dynamic safety capability

1. Understand the present state—what are our current risks and how are they controlled?
2. Any new information available? We need to look both inside our company and at any available, external sources.
3. Process the new information—how does it impact our current risk perception.
4. Act upon the new information. We need to consider both new risks and changes to our current set of risks.

When we get new information related to risks, it is important to remember Kristamuljana et al. (2018)—risks tend to cluster and are often correlated to other risks. Thus, if we get information that leads us to a new risk, we need to ask two important questions:

- Are any risks similar to this one that we have not considered yet?
- If the new risk is somehow related to a risk that we already are considering, should we update its probability or consequences?'

This is the same idea as is used in Coachio (2019)—scan and attend to future (ability to scan and interpret the external environment for oppurtunities and threats to safety), integrate complexity and second-order changes and also the same idea as the one presented by Denny et al. (2015)—monitor the system's environment and analyze, report, and identify.

As we see from Fig. 6.1, there are several ways to improve the safety when the environment or operation changes. The most common way to handle the new challenges will be to introduce new barriers in the software, the hardware, or in the equipment's operation. Note that a changing environment may also lead to the removal of one or more risks. In this case, the barriers might be removed—especially in software since software barriers add to the complexity and may make the system less reliable or safe.

6.11.3   Dynamic Safety Cases and DevOps

When the system's environment changes, we may need to also change the safety cases. We need to monitor both the system's environment and the way the system is used. Figure 6.11, adapted from Denny et al. (2015), shows the mechanisms at play for a dynamic safety case. Note that we here discuss how to update the safety case after the system has been put into operation—what we might call the "DevOps phase." However, the agile safety case process can still be used.

The agile safety case can be used as the process for developing the safety case and the operational safety case. How we identify new threats is special for the dynamic safety case.

It might be practical to make and maintain a Safety Case Report. This is simply a "snapshot" of the status of the safety case at a given point in time together with the current status of the arguments the evidence (see Kelly 2016). The safety case report

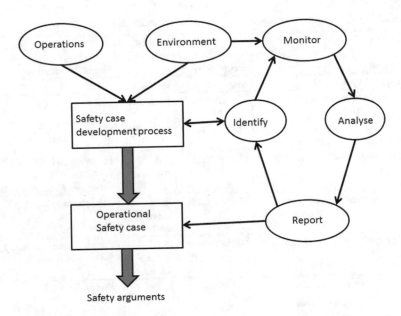

Fig. 6.11 The dynamic safety case

Fig. 6.12 Safety case life cycle

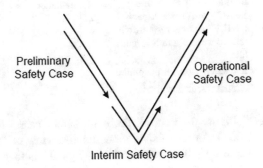

should be used, e.g., when using DevOps so that we can see which hazard analyses, arguments, and assumptions need to be updated. The safety case report will fit well together with the report shown in Fig. 6.4.

Defence Standard 00-55 formally issue three versions of the (Software) Safety Case:

- Preliminary Safety Case—after definition and review of the system requirements specification
- Interim Safety Case—after initial system design and preliminary validation activities
- Operational Safety Case—just prior to in-service use, including complete evidence of satisfaction of systems requirements. The integration between the production of these safety cases and the traditional development life cycle is depicted in Fig. 6.12.

Table 6.19 Safety cases—who prepare, approve, and authorize them

Safety case	Normally prepared by	Normally approved by	Normally authorized by
Concept safety case	Manufacturer	Internally	None
Building a preliminary safety case (Kelly et al. n.d.)	Manufacturer	Internally	None
Phase 4 (IEC 61508): Specification of system requirements safety case	Manufacturer	Internally	None
Generic product safety case	Manufacturer	ISA issues a SAR	In Europe, a NoBo or CB may issue a certificate for constituents and products
Generic application safety case	Manufacturer	ISA issues a SAR	Not relevant
Specific application safety case	Manufacturer	ISA issues a SAR	National safety authority in the railway domain
Cross acceptance safety case	Manufacturer	ISA issues a SAR	Not relevant
"Top" safety case. Includes evaluations of e.g., the SRAC's and similar aspects. If, e.g., the IM has some responsibility, the related work has to be argumented for.	Infrastructure manager	ISA issues a SAR	National safety authority in the railway domain
The safety case for the public (Myklebust et al. 2021)			

Figure 6.12 shows the main safety case life cycle actions according to Kelly (2003). An alternative, more detailed view of the safety case life cycle is shown in Table 6.19. The message from Fig. 6.12 is to start with a preliminary safety case—often based on reuse. As more information become available, the preliminary safety case is then extended and modified to become an interim safety case which in the end is made into an operational safety case.

The relationship between safety case development made during the software development process, and the operational safety case is shown in Fig. 6.13. The right-hand part of the activities described in Fig. 6.4 is also shown in the right-hand part of the diagram in Fig. 6.7.

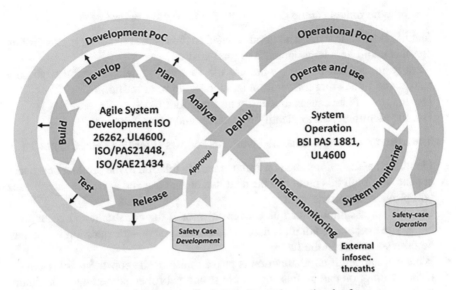

Fig. 6.13 The relationship between development safety and operational safety

6.11.4 Two Approaches to Analyzing Emergent Risks

There exist a large number of methods for analyzing emergent risks and hazards. We will present two simple methods below. For other, more advanced methods such as FRAM, you should consult the literature. Here we will just give a short presentation of the bow-tie model and the assumption analysis approach. These two models have some important things in common:

- They enable us to involve all relevant personnel and thus all relevant and available information.
- The process is simple, thus allowing everybody to participate and contribute.

Figure 6.14 shows the bow-tie model for the emerging trend "abundance of data" but the approach is general.

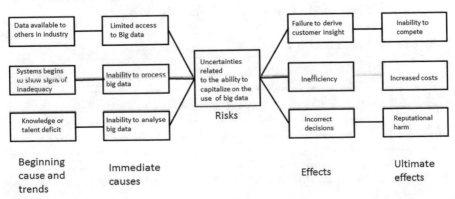

Fig. 6.14 Bow-tie model example

The bow-tie model can be described by the following steps:

- Identify beginning cause and trends related to the issue at hand—e.g., for driving car, this could be a large increase in road traffic.
- Identify immediate causes—what are the first consequences of the beginning causes and trends that follow from an increase in road traffic.
- Use the immediate causes to identify an important risk.
- Use the identified risk to identify immediate and ultimate effects.

The assumption analysis method works as follows (Fiorille et al. 2019):

- Identify all assumptions and constraints made in relation to the product or process at hand. For both this point and the next, the hazard log and available safety cases will provide important input.
- Check if an assumption can be proven false or if the constraint can be broken. Note: The question is not if the assumption is false now—the question is if it may be false sometime in the future.
- What can happen if an assumption is proven false or the constraint is broken?
- If the assumption can be false or the constraint is broken sometime in the future and may have a dangerous consequence, we have a new risk.

The new risk can be described using the following pattern:
<assumption / constraint> **may prove false, leading to** <effect on objects>.

6.11.5 Morphological Analysis: GMA

The GMA method is simple to explain but complex to use for the simple reason that the number of possible scenarios will grow exponentially. The basic idea can be shown in Table 6.20. The GMA table has one column per component and one line per state. Both the number of columns and the number of lines will depend on the chosen level of details in our analysis. In addition, the description and number of states will differ for each system component.

We can now construct scenarios based on the MA table—e.g., a scenario where component 1 is in state S1, component 2 is in state 5, and so on.

Table 6.20 The general GMA table

| State | System components | | | | | |
	Comp. 1	Comp. 2	. . .	. . .	Comp. n-1	Comp. n
S 1						
S 2						
S 3						
. . .						
. . .						
S m						

The problem is clear—we will have m^n possible scenarios. The common strategy is to use experts to identify impossible or unreasonable component–state combinations and ignore them for the rest of the process. One possible approach to use is the cross-consistency matrix. Other possible approaches are the subject matter expert or event trees. The only problem with this approach is that it may make us ignore possible "Black swans" lurking out there in the darkness.

Once we have a useful set of scenarios, we can start identifying defenses or barriers. There is a rich set of resilience patterns available, and we can choose the most appropriate ones based on experience and domain knowledge (see also Ritchey 2002).

References

AIAG & VDA: FMEA Handbook—Potential Failure Mode Effects Analysis, FMEA (2019)

Bao, H., Shorthill, T., Zhang, H.: Light Water Reactor Sustainability Program Redundancy-guided System-theoretic Hazard and Reliability Analysis of Safety related Digital Instrumentation and Control Systems in Nuclear Power Plants. U.S. Department of Energy Office of Nuclear Energy, August (2020)

Brancati, F.: A methodology to ensure safety (certification) of complex software in safety critical automotive systems. 75th Meeting of IFIP Working Group 10.4 Champery, Switzerland. 24–28 January (2019)

Brilliant, S.S., Knight, J.C., Leveson, N.C.: Analysis of faults in an N-Version Software Experiment. IEEE Trans. Soft. Eng. 16(2) (February 1990)

Denny, E., Habli, I., Pai, G.: Dynamic Safety Cases for Trough-Life Safety Assurance (2015). https://doi.org/10.1109/ICSE.2015.199

Fiorille, F., Graham, L., Kaufman, C.: Identifying and Evaluating Emerging Risks., Rims NZ (2019).

Griffin, M.A., Cordery, J., Soo, C.: Dynamic safety capability: How organizations proactively change core safety systems. Organizational Psychology Review. 6(3), 2015 (2016)

Halbert, M., Tucker, S.: Risk Assessment for M42 Active Traffic Management (2006). https://doi.org/10.1007/1-84628-447-3_2

Hauge, S., Hokstad, P., Håbrekke, S., Lundteigen, M.A.: Common cause failures in safety-instrumented systems: Using field experience from the petroleum industry. Reliab. Eng. Syst. Saf. 151 (2016)

ICDE Project Report: Collection and Analysis of Common-Cause Failures of Level Measurement Components. July 1, 2008

Kelly, T.: A Systematic Approach to Safety Case Management. SAE International (2003)

Kelly, T.: What does it mean to have a dynamic safety case? University of York (2016)

Kelly, T., Bate, I., McDermid, J., Burns, A.: Building a Preliminary Safety Case: An Example from Aerospace. (n.d.) www-users.cs.york.ac.uk/tpk/preliminary.pdf

Klim, H.Z., Balazinsky, M.: Methodology for Common Mode Analysis. SEA technical papers (2007)

Kristamuljana, A., van Loon, B., Bolt, J., Terblanché, A.: Dynamic risk assessment above and beyond the hidden structure of interconnections between risks, compact (2018/2) www.compact.nl/articles/dynamic-risk-assessment/

Leedeo: Common cause failures (CCF). What are they and how are they mitigated. 01/10/2020. www.leedeo.es/l/common-cause-failures-ccf/

Lindberg, S.: Common cause failure analysis Methodology evaluation using Nordic experience data. Uppsala University (May 2007)

Malm, T., Hietikko, M.: Safety Validation by VTT for SAFELOC Round Robin Tests of Safety-related Control System for Machinery. NordTest, 31th of October 2000

MIL-HDBK-217F: Notice 2 parts count reliability prediction pages (2018)

Myklebust, T., Onshus, T., Lindskog, S., Ottermo, M.V., Bodsberg, L.: Data Safety, Sources, and Data Flow in the Offshore Industry. ESREL, Angers (2021)

Papadopoulos, Y., McDermid, J.A.: Hierarchically performed hazard origin and propagation studies. Comput. Saf. Reliab. Sec. Lect. Notes Comput. Sci. **1698**, 139–152 (1999)

Puisa, R., Bolbot, V., Ihle, I.: Development of functional safety requirements for DP—driven servicing of wind turbines European STAMP Workshop & Conference, 2019 at Aalto University, Finland

Raman, R.: Process systems risk management. Chapter 4: Identifying hazards and operational problems (2005)

Rausand, M., Høyland, A.: System Reliability Theory. Models, Statisitcal Methods and Applications, 2nd edn. Wiley, Hoboken, NJ (2004)

Ritchey, T.: General Morphological Analysis—A general method for non-quantified modelling. Swedish Morphological Society 2002 (revised 2013)

Ross, D.: Risk, The Application of As Low as Reasonable Practicable (ALARP), August 19, 2019 rmstudy :Identify risks, http://rmstudy.com

SCSC: Data Safety Guidance. https://scsc.uk/SCSC-127E (2021)

Squair, M.: Hazard Checklists, and Their use in Hazard Identification. Checklist version 1.2 (2014)

Stålhane, T.: FMEA, HazId, and Ontologies in Ontology Modeling in Physical Asset Integrity Management. Editors: Ebrahimipour, Vahid, and Yacout, Soumaya (2015)

Stålhane, T., Malm, T.: Four Perspectives on Safety Analysis. ESREL, Venezia (2020)

The COACHIO Group: Dynamic Risk Assessment and Management (2019). www.healthandsafety.govt.nz/reports/presentations/dynamic-risk-assessment-and-management/

Thomson, J.: Common-Mode Failure Considerations in High-Integrity C&I Systems (February 2012)

Zollo, M., Winter, S.G.: Deliberate learning and the evolution of dynamic capabilities. Org. Sci. **13** (2002)

Chapter 7
Safety and Risk Documents

*Men ought therefore to look to the risks and dangers of any
course which lies before them, nor engage in it when it is
plain that the dangers outweigh the advantages, even though
they be advised by others that it is the most expedient way
to take.*

*Niccolò Machiavelli: Discorsi sopra la prima deca di Tito
Livio*

What This Chapter Is About
- Hazard logs and their relations to safety standards
- Safety analysis report
- Hazard and risk analysis report

7.1 Hazard Log

The hazard log (also named hazard record) lists and tracks all hazards, hazard
analysis results, risk assessments, and risk reduction activities during the whole
lifetime of a safety-related system. A complete hazard log is normally one of the top
five references in a safety case. The agile hazard log approach enables the manufacturer to have a single source for risk management activities and simplifies the reuse
and transfer of information between projects.

The benefits of using a hazard log are that:

- It is one of the main references in the safety case. A safety case is required by,
  e.g., EN 50129:2018 (railway), ISO 26262:2018 (automotive), and UL
  4600:2020 (automotive).
- The European railway regulation for "the common safety method for risk evaluation and assessment" (402/2013 and EU 2015/1136) requires a hazard log
  (hazard record) to be created.
- It presents an overview of the hazards and, e.g., the relevant responsibility and
  mitigations.

© The Author(s), under exclusive license to Springer Nature Switzerland AG 2021   163
T. Myklebust, T. Stålhane, *Functional Safety and Proof of Compliance*,
https://doi.org/10.1007/978-3-030-86152-0_7

EN 50126-1:2017 and EN 50129:2018 define a hazard log as "document in which hazards identified, decisions made, solutions adopted, and their implementation status are recorded or referenced".

Our suggestion for a definition of an Agile hazard log is:

> The Agile hazard log: Information on all safety management activities, hazards identified, decisions made, and solutions adopted are recorded. This should be collected and registered in an adaptive, flexible, and effective way.

Companies introducing agile methods like SafeScrum should also use an Agile hazard log (AHL) to get the full benefit of an agile approach and at the same time satisfy relevant safety standards like, e.g., EN 5012X series, ISO 26262 series, and IEC 61508 series. The main reason for introducing and applying the AHL is that:

- It ensures a more complete agile approach, ensuring an agile approach also for the RAM/safety engineering team.
- The AHL process is an improved process for handling and mitigating hazards, especially regarding the stronger emphasis on software development.
- The AHL process is suited for frequent updates due to process improvements.
- The AHL may facilitate a single-source approach for risk management activities.
- It may simplify the reuse and transfer of information between stakeholders, depending on the tool used.

Companies or organizations having several and/or big projects should consider establishing a company hazard log to ensure:

1. Establishing possible scenarios and possible mitigations
2. Standardizing hazard case descriptions to avoid repetition
3. Collecting data on hazards to create an open-source hazard log
4. Enabling delta analysis for working with new projects

The introduction of the AHL ensures focus on hazards at, e.g., the daily scrum meetings, the sprint reviews, and the AHL's focus in general. A hazard log that is not adapted to frequent changes may quickly become outdated, in the sense that it no longer represents the true picture of the risks related to the product being developed by the manufacturer. Using an AHL enables a structured, agile, and flexible approach allowing for frequent software updates and a shorter time to market. One of the main references in the safety case, the AHL, is up to date with the current safety picture.

The introduction of the AHL contributes to avoiding SW design errors as the AHL is part of the Sprint reviews. Current standards are weak and do not match the current and future heavy focus on software processes. A hazard log that is not well adapted to frequent changes may easily become outdated, in the sense that it no longer represents the true picture of risks related to the product being developed by the manufacturer. The AHL enables a structured, agile, and flexible approach allowing for more frequent updates and a shorter time to market.

The RAMS team develops the AHL alongside the product development—i.e., in activities performed alongside the sprints. Sprint is a Scrum term used to describe an iteration of the development in the Scrum process. The AHL-related work can be time boxed together with the Sprints but is performed alongside the Sprints. The Sprint review may include the AHL as a topic when relevant. The development of the AHL should preferably be planned together with other safety activities like the development of the agile safety case.

Further, the requirements for the hazard log and the AHL are presented in the chapter below. To ensure an efficient project, we have included the reuse and template approach. Finally, we present the agile hazard log process and relevant agile practices.

The rest of this section is organized as follows: a requirement chapter and possibilities related to the reuse of templates. We have not included the following phases in the current process description: manufacturing, installation and commissioning, operation and maintenance, and decommissioning.

The definitions below are based on definitions from the relevant generic standards like ISO/IEC Guide 51:2014, IEC 60050-821:2017, ISO 31000:2018, and safety standards. The CENELEC standards include definitions in each EN 5012X. Harm, hazard, hazardous event, and hazardous situation are not defined in EN 50128:2011. In the two tables below, we have shown the definitions for some of the most important terms (Tables 7.1 and 7.2):

7.1.1 HL and Requirements in Safety Standards

The Agile HL must satisfy the relevant safety standards, so the requirements for an Agile HL and an ordinary HL are more or less the same.

7.1.1.1 Generic Safety Standard

The generic standard series IEC 61508 does not have requirements for a HL but includes as part of phase 3 requirements for hazard and risk analysis (see Fig. 7.1). IEC 61508-1 includes three objectives: "The first objective of the requirements of this subclause is to determine the hazards, hazardous events and hazardous situations relating to the EUC (Equipment Under Control) and the EUC control system (in all modes of operation) for all reasonably foreseeable circumstances, including fault conditions and reasonably foreseeable misuse.

The second objective of the requirements of this subclause is to determine the event sequences leading to the hazardous events.

The third objective of the requirements of this subclause is to determine the EUC risks associated with the hazardous events."

These objectives agree well with the EN 50126-1 requirements for a HL. Therefore, even though not required by IEC 61508, the required hazard and risk analyses may be included or referenced in a HL.

Table 7.1 Definitions of harm and hazard

Standard term	IEC 61508-4:2010	ISO 26262-1:2018	EN 50126-1:2017	EN 50129:2018
Harm	3.1.1 Harm Physical injury or damage to the health of people or damage to property or the environment ISO/IEC Guide 51:1999, definition 3.3	3.74 Harm Physical injury or damage to the health of persons	Not defined but used PoC note: See also explanation as part of the definition below	Not defined but used PoC note: See also explanation as part of the definition below
Hazard	3.1.2 Hazard Potential source of harm ISO/IEC guide 51:1999, definition 3.5 Note: The term includes danger to persons arising within a short timescale (e.g., fire and explosion) and also those that have a long-term effect on a person's health (e.g., release of a toxic substance)	3.75 Hazard Potential source of harm caused by malfunctioning behavior of the item Note 1 to entry: This definition is restricted to the scope of the ISO 26262 series of standards; a more general definition is potential source of harm	3.28 Hazard Condition that could lead to an accident Note 1 to entry: The equivalent definition in IEC 60050-903:2013, 903-01-02 refers to "harm" instead of "accident"	3.1.21 Hazard, <in railway> Condition that could lead to an accident Note 1 to entry: The equivalent definition in IEC 60050-903:2013, 903-01-02 refers to "harm" that, with respect to "accident," does not include loss of system or service

7.1.1.2 Railway

In the EN 5012X series, the main requirements related to hazard identification and hazard processes are included in EN 50126-1:2017 and EN 50129:2018, while the requirements and information in EN 50128 are of little help except for the requirement that the validator has to *ensure that the related hazard logs and remaining non-conformities are reviewed and that all hazards are closed in an appropriate manner through elimination or risk control/transfer measures.*

The majority of requirements related to a HL in EN 50126-1:2017 and EN 50129:2018 are on the HL itself and, to a lesser degree, specific requirements on the process, even though EN 50126-1:2017 states when in which lifecycle phases the HL shall be updated or reviewed.

The European railway regulation for "the common safety method for risk evaluation and assessment" (402/2013 and EU 2015/1136), hereafter denoted CSM, requires a hazard log (hazard record). Whereas EN 50126-1:1999 presents requirements to the HL on a rather detailed level, CSM presents the requirements on an

Table 7.2 Definition of hazardous event and hazardous situation

Standard/ term	Hazardous event	Hazardous situation
ISO/IEC guide 51:2014	3.3 Hazardous event Event that can cause harm	3.4 Hazardous situation Circumstance in which people, property, or the environment is exposed to one or more hazards
IEC 61508-4:2010	3.1.4 Hazardous event Event that may result in harm Note: Whether or not a hazardous event results in harm depends on whether people, property, or the environment is exposed to the consequence of the hazardous event and, in the case of harm to people, whether any such exposed people can escape the consequences of the event after it has occurred	3.1.3 Hazardous situation Circumstance in which people, property, or the environment is exposed to one or more hazards ISO/IEC Guide 51:1999, definition 3.6, modified
ISO 26262-1:2018	3.77 Hazardous event Combination of a hazard and an operational situation	Not defined
EN 50126-1:2017	Not defined but used and includes a table: Table C.1—Frequency of occurrence of hazardous events with examples for quantification (time based)	Not defined and not used
EN 50129:2018	Not defined and not used	Not defined but used once

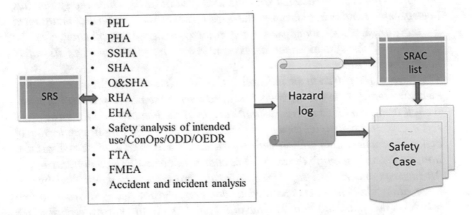

Fig. 7.1 From SRS, hazard analysis to the hazard log

overall, generic level. The reason for this may be that CSM defines three approaches (application of codes of practice, comparison with similar systems, or explicit risk estimation) for risk acceptance that may put different constraints on the HL. Important aspects to highlight for HL requirements are that CSM requires that the HL is created or updated by the proposer (EU terminology) of the change, e.g., the manufacture or the infrastructure manager, during design and implementation until acceptance of the change or delivery of the safety assessment report. Once the system has been accepted and is in operation, the HL shall be maintained by the infrastructure manager or the railway undertaking in charge of the operation of the system under assessment. CSM, therefore, requires that the responsibility is transferred between the proposer (responsible for the design/change) and the actors responsible for the operation of the system. Other aspects that should be highlighted related to the HL are that CSM presents concrete requirements to the system definition, determining the scope of the risk assessment, and the HL. This is also the case for EN 50126-1. CSM further emphasizes that the expertise from a competent team shall be used as part of the hazard identification. The risk management process is also, according to CSM, iterative and ends when compliance with the safety requirements necessary to accept the risk is demonstrated.

The AHL has to satisfy the relevant safety standards. Thus, the requirements for an AHL and an ordinary HL are more or less the same. It is the processes that are different. For the AHL, it is recommended to include all the hazard log requirements from the safety standard EN 50126 and communication aspects included in the CSM regulation, which also corresponds perfectly with the Agile mindset.

Compared to EN 50126-1:2017, CSM regulation presents the HL (hazard record) requirements on a generic level. It states that:

- *Hazard record(s) shall be created (or updated where they already exist) by the proposer during design and implementation until acceptance of the change or delivery of the safety assessment report. A hazard record shall track the progress by monitoring risks associated with the identified hazards. Once the system has been accepted and is in operation, the hazard record shall be further maintained by the infrastructure manager or the railway undertaking in charge of the operation of the system under assessment as an integrated part of its safety management system.*
- *The hazard record shall include all hazards, together with all related safety measures and system assumptions identified during the risk assessment process. It shall contain a clear reference to the origin of the hazards and to the selected risk acceptance principles (three principles in the CSM regulation: application of codes of practice, comparison with reference systems, and explicit risk estimation) and clearly identify the actor(s) in charge of controlling each hazard.*
- *All hazards and related safety requirements that cannot be controlled by one actor alone shall be communicated to another relevant actor in order to find jointly an adequate solution. The hazards registered in the hazard record of the actor who transfers them shall only be regarded as controlled when the evaluation of the risks associated with these hazards is made by the other actor and the solution is agreed by all concerned.*

When comparing the requirements to the HL from EN 50126-1 and CSM, as presented above, the communication aspects when several actors are involved and the responsibilities of the HL are to a larger degree covered by CSM than EN 50126-1.

7.1.1.3   Automotive

The automotive safety standard ISO 26262 does not have requirements for a HL but includes it as part of ISO 26262-3 hazard analysis and risk as part of the concept phase and other parts concerning safety analysis.

The AHL is developed alongside the product/system and is normally maintained by the safety/alongside engineering team—i.e., in activities performed alongside the sprints. The AHL-related work can be time boxed together with the sprints but is performed alongside the sprints. The sprint review may include the AHL as a topic when relevant. Updating the AHL should preferably be planned together with other alongside sprint activities like the development of the ASC, analysis, and independent tests. The AHL has to satisfy the requirements in EN 50126-1:2017 and EN 50129:2018.

The hazard log is a living document (see Sect. 7.1.1) or information system. If having a small project, it can be sufficient to use Word or Excel to handle the hazard log, but using these tools for large projects is not recommended. There can be many hazards in a project. Vinerbi and Puthuparambil (2017) presents different metro systems that include from 1400 to 2500 hazards.

A hazard log should be easy to use and be auditable. Some tools exist for hazard logs, and some integrate the hazard log in tools used for other purposes as, e.g., the SRS. The figure below shows the information flow from and to the hazard log.

Several techniques can be used by the manufacturer when performing hazard identification. For example, we may use historical safety experience and hazard logs, lessons learned, trouble reports, hazard analyses, and accident and incident files. When an acceptable edition of the "Definition of system" and the ODD exist (operational envelope), the hazard identification should start. The system definition is an important document for developing both the hazard log and the safety case. The system definition is also of crucial importance for the production of evidence to be included in the safety case. With a frequently changing world, the system, and thus the system's definition, has to be changed frequently. One way to handle this is to change from a waterfall process to an agile process.

7.1.2   Input Documents and Related Plans

Below are the relevant input documents listed together with related plans (Table 7.3).

Table 7.3 Relevant input documents and related plans

Input documents	Related plans and documents
• Relevant safety standards, e.g., IEC 61508, ISO 26262, and EN 5012x series • Description of the system (DoS) • ConOps, intended use, description of the ODD including OEDR • Relevant HazId information—See figure above • Railway UNISIG subset 113 hazard log • Railway CSM regulation: EU No 402/2013 and EU 2015/1136	• Project plan • HAZOP plan and related plans if not included in the safety plan • Safety plan issued by suppliers • Subcontractor hazard log • Product hazard log • Operators hazard log

7.1.3 Recommended Process Approach

EN 50126-1 is one of the few international safety standards requiring a hazard log being developed by the manufacturer (and the operator). Therefore, it is natural to investigate some of the overall requirements to the HL in this standard to provide parts of the basis for the development of the AHL. According to EN 50126-1: 2017, the HL "shall be updated throughout the life-cycle whenever a change to identified hazards occurs or a new hazard is identified." This implies that the HL may be updated in any of the lifecycle phases. However, the establishment of the hazard log is to be performed as part of *Phase 3 Risk analysis*. This can be seen from EN 50126-1 because most of the detailed requirements to the HL are contained in the clauses of EN 50126-1 related to *Phase 3 Risk analysis*.

Even though *Phase 3 Risk analysis* is the phase where the HL is first established according to EN 50126-1, some of the foundation and prerequisites for establishing the hazard log are made during *Phase 2: System definitions and application condition* by defining the scope of system hazard analysis and by performing preliminary hazard identification.

The automotive safety standard ISO 26262-3:2018 correctly includes hazard analysis as part of the concept phase.

In addition to the railway safety lifecycle *Phase 3 Risk analysis* (the phase when HL is established), the HL is mentioned with respect to updates or reviews as part of lifecycle *Phase 10: System acceptance, Phase 11: Operation and maintenance, Phase 13: Modification and retrofit*, and *Phase 14: Decommissioning and disposal*.

The figure below includes the relevant safety lifecycles, including the main part of SafeScrum and relevant hazard analysis and agile practices. Safety stories—user stories that include one or more safety requirements—have been described earlier by both Myklebust and Stålhane (2016) and Paige (2011). Safety stories are a safety practice developed to ensure that the agile requirements management process encompasses the safety requirements and required measurements and techniques. In addition, the safety story will create a common problem understanding between

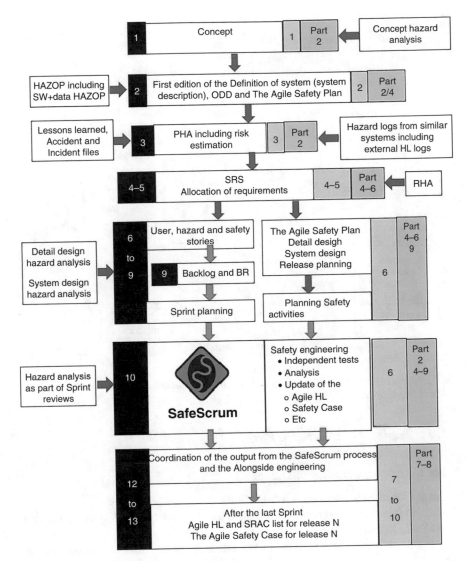

Fig. 7.2 Relevant lifecycle elements for a safety case

the software developers and the safety stakeholders. The Sprint team store the safety story in the safety part of the product backlog (Fig. 7.2).

Below, we have shown a sprint including four questions and sprint review that includes agile hazard log (AHL) discussions (Fig. 7.3).

Below, we have listed the topics relevant to a safety plan and general information and relevant agile adaptations.

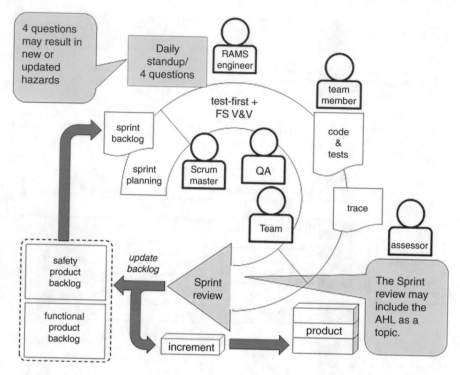

Fig. 7.3 A sprint including four questions and hazard log as part of the sprint review. Variant based on Fig. 4.3 in Hanssen et al. (2018)

7.1.4   Topics 1–7: Generic Parts

Topic 1 The purpose of the hazard log.

General Comments and Agile Adaptations
Domain-independent chapter topic but used to adapt the hazard log to the domain, project, and intended use.

Purpose example: The purpose of the HL is to ensure that all the hazards are mitigated and that none of the hazards results in SRACs that affect the operation and maintenance of the product.

Purpose (agile): This product shall be developed having a design that ensures a limited number (<10) of SRACs are transferred to the IM. The remaining SRACs will be solved by upgrading the SW in the next three releases.

Topic 2 Hazards, entities responsible for managing the hazard, and the contributing functions or components.

General Comments and Agile Adaptations
Domain-independent chapter topic but used to adapt the hazard log to the domain, project, and intended use.

Table 7.4 Top hazards for the railway, hyperloop, and automotive

Railway	Hyperloop	Automotive
• Derailment • Collision train-train • Collision train-object • Fire • Passenger falling from a platform or similar • Level crossing collisions	• Collision pod-pod • Fire • Pressure loss • Thermal expansion • Level crossing collisions	• Collision car-car • Collision car-object • Fire

Often, there exist only a few *top hazards* (typically three to ten top hazards in the railway signalling domain). Top hazards could preferably, for some companies, benefit being included in an epic.

The limits of any analysis carried out should be referred to or described. It is potentially challenging to determine the limits of analyses and scope of that hazard log when the system is complex and when there are many actors involved. Each actor may have their own responsibilities in terms of the development of the system (e.g., different actors developing different parts of the system) and the operational aspects of the system. There may be several HLs that need to interact in certain cases, e.g., the different actors may each control their own HL.

As part of the sprint team, the QA role (Hanssen et al. 2016) could, e.g., be responsible for the communication of the hazard log within the Sprint team. When using a DevOps approach (which often involves more stakeholders), the project manager should ensure that the entities responsibility is clarified early in the project (Table 7.4).

Railway: Copy from subset 091 ed. 2016: 4.2.1.8 Thus, it is necessary to define two different hazards of ETCS to distinguish between these two situations:

- For the case that ETCS has information on safe speed and distance (hazard is denoted "ETCS Core Hazard")
- Exceeding the safe speed or distance as advised to ETCS
- For the case that ETCS does not have information on safe speed and distance (hazard is denoted "ETCS Auxiliary Hazard")

Topic 3 Likely consequences and frequencies of the sequence of events associated with each hazard, when applicable.

General Comments and Agile Adaptations
Domain-independent chapter topic, but it is important to adapt the hazard log to the domain, project, and intended use.

Different approaches exist, e.g., detailed causal analysis including fault tree analysis (FTA) calculation to determine frequencies and consequences for the sequence of events (causes) associated with each hazard.

- Engineering judgement of the consequence and frequency of the sequence events associated with each hazard. Events contributing to FTA can be determined by the use of, e.g., the Agile FMEA approach (Myklebust et al. 2018).

- Concerning FTA, several approaches exist, e.g.:
  - Detailed calculation by fault tree analysis to determine frequencies (and consequences) for the sequence of events (causes) associated with each hazard
  - Engineering judgement of the consequence and frequency of the sequence events associated with each hazard

For railway, see also UNISIG Subset-077 "Causal analysis" and UNISIG subset 113 ETCS hazard log.

Topic 4 The risk arising from each hazard (in quantitative or qualitative terms), where appropriate.

General Comments and Agile Adaptations
Risk is defined as the product of the frequency or probability and the consequence of a specified hazardous event. These are often presented in a risk matrix. The work related to the high risks should be prioritized at the sprint planning meetings.

Topic 5 *Risk acceptance principles selected and, in the case of explicit risk estimation, also the risk acceptance criteria to demonstrate the acceptability of the risk control related to the hazards.*

General Comments and Agile Adaptations
The risk tolerability is not defined by the standard. Tolerable risk can be determined by:

- Goals and or strategy by the manufacturer
- The current values of society
- The search for a balance between the ideal of safety and what is achievable
- The demands to be met by a product or system
- Factors such as suitability for purpose and cost-effectiveness

The tolerable risk may be defined by, e.g., the safety authorities or within the railway domain, the IM.

Railway: See also RSSB GE/GN 8643 (2017) Guidance on risk evaluation and risk acceptance.

Oil and gas: See NOG 070 (2020) for different SIL for the different products/ systems.

Topic 6 *For each hazard: document the measures taken to reduce risks to a tolerable level or to remove the risks. This part should include relevant arguments.*

General Comments and Agile Adaptations
This subject is sometimes also discussed in meetings between the manufacturer and the operator. Operators may control several hazardous events by including instructions in manuals, operational rules, traffic rules, etc. A potential challenge may be how to ensure that future changes within manuals, operational rules, or traffic rules do not negatively affect risks related to the hazards in the HL. This could, e.g., be part of the Sprint review meetings.

Topic 7 *Exported safety constraints.*

General Comments and Agile Adaptations

Exported constraints from, e.g., GPSC to SASC or, e.g., subcontractor to integrator could, e.g., be as an open hazard log item or as a SRA. Templates and example of a simple hazard log.

In the table below, we have presented an example of a simple hazard log (Table 7.5):

In the template below, we have included a template for how a new hazard can be reported (Table 7.6):

The Agile HL approach takes the best from the two worlds: conforming to safety standards and agile processes and practices. We have presented recommendations for the AHL requirements: it is recommended to include all the hazard log requirements from the safety standard EN 50126-1 together with communication aspects included in the CSM regulation, which also corresponds to the agile mindset. A safety process has been suggested involving, among other things, safety stories, SafeScrum, and alongside engineering. Adaptation of the agile practices may result in further improvements of the process: backlog together with backlog splitting, sprint review, and backlog refinement.

The AHL approach further represents an improved process, e.g., by being suited for frequent changes to develop and maintain the HL. The process is especially improved in relation to the stronger emphasis on software development.

This work has also shown that there exist possible improvements of IEC 61508-3, by requiring the use of hazard log as a single source of risk management activities. This task, performed by the validator, should preferably also be required in the IEC 61508 series without requiring who shall perform the task: "the validator has to ensure that the related hazard logs and remaining non-conformities are reviewed and that all hazards are closed in an appropriate manner through elimination or risks control/transfer measure."

Table 7.5 Hazard log example

Hazard identity		Hazard description			Risk evaluation (before reduction)			Risk reduction measure		Status		Notes and comments
ID	Reference	Description	Hazard cause	Type of accident	Severity	Frequency	Risk	Measure/function	Responsible	Reference	Status	
I1	A1	Driver controllability	Driver	Collision	Catastrophic	Infrequently	Intolerable	Improved training and operational manuals	*** ********		Open	

Table 7.6 Hazard item template

Hazard item			
System, product, ODD/OEDR, intended use, operation			
Hazard identification/number			
Hazard description (causes, sources, threats)			
Potential outcomes Including consequences and magnitudes			
Risk controls (barriers and mitigations)			
No.	Description		Responsible
1			
2			
3			
4			
5			
Risk assessment			
Hazard frequency			
Outcome likelihood			
Consequence severity			
Risk			
Safety manager approval	Name:	Date:	Signature:

7.2   Safety and DevOps

7.2.1   Introduction

DevOps differs from ordinary software development in that its target area is to give end-to-end business solutions and fast delivery. The main characteristics are that DevOps:

- Brings development and operations teams together
- Focuses on constant testing (automatic) and delivery
- Requires a relatively large team since there are several parties involved:

  - Manufacturer (several teams)
  - Certification and approval bodies
  - Operator(s)
  - Users

- Leverages both shift-left and shift-right principles
- Focuses more on operational and business readiness (Fig. 7.4)

The differences between shift-right testing and shift-left testing can be described as follows:

- Shift-left testing—Start testing earlier.
    The test team starts performing tests earlier in the development cycle. Finding and fixing bugs as early as possible helps you introduce higher-quality code right from the beginning and save time and resources.

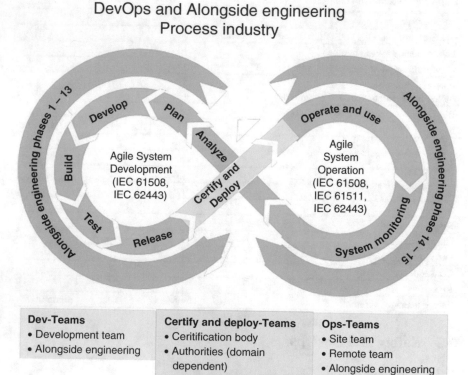

Fig. 7.4 DevOps and alongside engineering

- Shift-right testing—Post-deployment testing.

   By employing shift-right testing, you can learn how your product operates under real-world conditions and how actual users handle it.

A DevOps process usually starts with an error report or a complaint from a customer. The complaint may often concern changes to an existing function or the request for a new one—all included under the term "maintenance." We will consider four types of maintenance activities as shown below:

- Adaptive maintenance (18%) is implementing changes in a part of the system, which has been affected by a change that occurred in some other part of the system.
- Perfective maintenance (60%) deals with implementing new or changed user requirements.
- Corrective maintenance (17%) is the implementation of error corrections—bug fixing.

- Preventive maintenance (5%) involves performing activities to prevent the occurrence of errors. This is not a part of DevOps and will thus not be discussed any further.

The percentages are related to person-hours used. Note that this will vary depending on type of system and type of customers. The data used here are taken from Schach (2003). All four categories are important for DevOps and the relationship between DevOps and safety.

7.2.2   DevOps vs. Maintenance

What separates DevOps from regular maintenance is the degree of customer involvement—both as a provider of test cases and their place in the loop—remember DevOps brings developers and operators together since both parties contribute test cases. The challenge is the large number of changes that can be introduced by working this way. Changes may create problems for two reasons:

- They may introduce new errors or bugs.
- They require a change impact analysis and a new safety analysis.

Since DevOps will cause many changes to and from, we run the risk of having a large number of safety analyses, which can be costly. However, there is also an upside. The test data supplied by the operator is real-world data.

DevOps will involve the operators. If they have little or no experience with safety analysis, it might be difficult to involve them except after a large investment in training. On the other hand, since they use the system in a real environment, they may give important input about safety-related events, near misses, how they occur, and why they are important to deal with (Fig. 7.5).

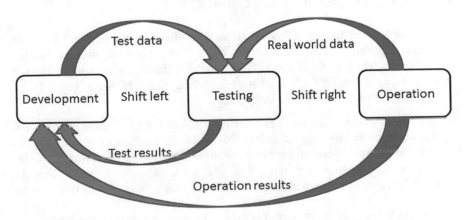

Fig. 7.5 DevOps combines shift-left and shift-right testing

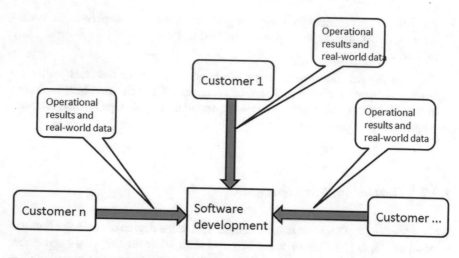

Fig. 7.6 DevOps combing shift-left and shift-right testing with several customers

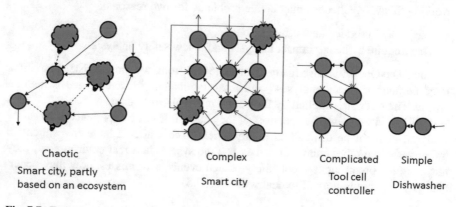

Fig. 7.7 Examples of systems with varying complexity

Another problem related to DevOps is that a development company probably will have many customers. Thus, there will be many changed or new requirements, and they will all most likely affect safety. Thus, instead of the nice, clean situation shown in Fig. 7.6, we will most likely have something like what is shown in Fig. 7.7.

There are at least two ways to handle this situation:

- We can coordinate all change requests. For safety-critical systems, the updates must be followed by re-certification. This will limit the number of new versions released per year to something between one and six for practical reasons. Coordination, combined with prioritization, will work satisfactory even though the need for re-certification will make the whole process much slower than it is for "ordinary" DevOps.
- We can let each customer have his own version of the system. Combined with close cooperation with the certification authorities, this will take care of the "need

for speed" but will, over time, make the maintenance and DevOps processes almost impossible due to a strong proliferation of versions to maintain. This proliferation of versions should be taken care of by using configurations instead of versions.

It is also possible to start with the second alternative and then merge the development branches after delivery. This will, however, add extra cost but will enable us to come up with a consolidated version after some time. Your preferred choice of action will depend on the number of maintenance requests.

7.3 Safety Analysis Reports

7.3.1 Introduction

The structure of this section is based on part 6.5 of IEC 61882:2016 "HAZOP." According to IEC 61882, there are two ways to report the results of a safety analysis: full recording—recording all results—and by exception, recording only the identified problems and the follow-up actions. For the sake of report completeness, we will record all results. Thus, this section will have the following contents:

- Safety threats and operability problems and how to mitigate them.
- Recommendations for further studies of specific aspects of the design.
- How to address uncertainties discovered during the study.
- Recommendations for mitigation of uncertainties related to the safety threats identified, based on the team's knowledge of the system.
- Points that need to be considered during operating and maintenance.
- The persons participating in the analysis.
- A list of the parts analyzed. If any have been excluded, the report should explain why.
- A list of the information used in the analysis.

As all other PoC reports, the safety analysis report should have a front page in line with the template shown in Sect. 1.2 Overview of documents, information, and work products.

7.3.2 The Analysis Report

This section gives a description of the structure and contents of the safety analysis report. The report's main purpose is to show the recipient how safe the system is, where it lacks safety, and what might be done about it. Therefore, the report format should be independent of the methods and methodology used. Assumptions and limitations of the analysis are handled in Chap. 6 "Safety Analysis Methods Applied to Software."

A company needs to have a standard format for the safety analysis report. In this way, the analysis team will know what information should go into the report, and they thus need an activity to make sure that this information is available. In addition, the level of details in the report will depend on when in the process (lifecycle) the analysis is done. For example, a safety analysis report from a HazId done at the concept level will be different from a safety report from the analysis of a finished system.

Last, but not least, we need to consider the complexity of the system. For a simple system, using a single safety analysis method will most likely be sufficient. However, for more complex systems, we need to use more methods in order to cover the full complexity.

Some examples of systems with varying complexity are shown in the figure above. Standard approaches such as FTA, FMEA, or HAZOP will most likely be sufficient for simple and complicated systems, but they will probably not suffice for complex and chaotic ones.

7.3.2.1   Safety Threats and How to Mitigate Them

The first thing to include here is references to the results from the analysis—e.g., FMEA tables, HAZOP tables, or fault tree diagrams. Since the methods will be used at different points in a system's development, the mitigation suggestions may also vary accordingly. Mitigations may include but are not limited to:

* Design changes—mainly during early phases
* Insertion of barriers
* Improved process—e.g., additional tools or methods
* Operator training and instructions

7.3.2.2   Recommendations for Further Studies

The need for further studies should be seen together with issue 3 on uncertainties. Further studies may include:

* The same system analyzed using a different method
* Analysis of parts of the system that was not included by the manufacturer in the reported study

7.3.2.3   How to Address Uncertainties

There are several ways a safety analysis can carry uncertainties. Some of the important ones are:

* Are the failures independent? Interdependencies may change our assessment of probabilities and consequences. There are checklists that can be used to assess dependencies—at least a common mode analysis, see Klim and Balazinsky (2007).

- Are there unknown or ignored environmental factors? Unknown environmental factors may influence consequences and probabilities. To handle this, we should include the system's future operators and the persons who shall install and maintain it.
- Events with low probability but with large consequences—often "black swans."

GAM—General Morphological Analysis—is a general method that can be used to create scenarios that will be of great help in discovering situations not covered in the safety analysis; see Ritchey (2002). Some companies also use the tables in Annex A of IEC 61508 as a set of checklists to make sure that all required process activities have been performed.

For safety analysis, it is quite common to assume a stable situation. However, in several real-life situations, the environment may change over time, thus creating a dynamic risk situation. According to IHASCO (2021), a dynamic risk assessment is a continuous process of observing, assessing, and analyzing an environment to identify hazards and remove risk while completing a task. They require some quick decisions about safety. They *don't replace a risk assessment* but can complement one when you need to assess any unknowns that cannot be predicted or that change during your task. They allow for flexibility, continual assessment, and changing environments. It was initially introduced for fire services but has been extended to situations which have the potential to evolve.

Since we need to make assumptions on the environment in safety analysis, uncertainty in environmental knowledge will create uncertainty in safety analysis results. Thus, possible future changes to the system's environment should be documented together with their possible influence on the safety analysis results.

7.3.2.4   Recommendations for Mitigation of Uncertainties Related to Safety Threats

For a complex system, we can never be sure that we have discovered all challenges. Thus, one or more resilience mechanisms should be considered. A good description of how to achieve resilience in software-intensive systems is given in Friedrichsen (2016).

7.3.2.5   Points that Need to be Considered during Operation and Maintenance

This issue is especially related to system maintenance. Most system safety analyses are based on one or more assumptions. When doing maintenance, we need to ensure (1) that all assumptions also hold after the maintenance actions or (2) that the necessary safety analysis actions are performed based on the new set of assumptions.

7.3.2.6 The Persons Participating in the Analysis

This part of the report must include the following formation for all personnel that have participated in the analysis or provided information that has been used:

- Contact information
- Relevant educations
- Relevant training
- Relevant experience

This also holds for persons involved with giving information, e.g., through brainstorming, focus groups, or interviews. It is especially important to use this information to see which areas of expertise have been included and which areas are missing.

7.3.2.7 A List of the Parts Analyzed

Which parts of the system have been included in the analysis and—more importantly—what has not been included? An important issue to consider is: What is part of the system and what is outside of it? An additional challenge is the level of details. For example, there is a large difference between the two statements (1) we have analyzed subsystem X and (2) we have analyzed all subroutines in system X. Thus, a description of the level of details for the analysis is an important part of the safety analysis report.

7.3.2.8 A List of the Information Used

All information used must be identified. This includes drawings, diagrams, requirements, environment descriptions, and articles used as support material during analysis. If we have performed interviews or used focus groups to obtain information, e.g., about using the system, the generated material must be included in the list. Available training manuals and manuals for maintenance should also be included here.

7.4 Hazard and Risk Analysis Report

7.4.1 Introduction

The structure of this chapter is based on the requirements of IEC 60812:2018. We have also looked at ISO 26262 and IEC 61508, but neither has a good table of contents for hazard and risk analysis reports even though ISO 26262 has requirements for a report on how the developers have arrived at the requirements for the parameters of the ASIL requirements matrix—see ISO 26262-2, clause 6.4.11.4.

According to IEC 60812, the report should include the following 11 issues:

1. A description of the system
2. A clear description of scope and boundaries
3. Assumptions made
4. A clear, detailed description of the methodology used
5. Identification of stakeholders
6. A description of the method used
7. Sources of data
8. Identification of failure modes, guidewords, or other terms used
9. A summary of the results and recommended treatments
10. Limitations
11. Analysis records

We will discuss these requirements in some more details in the following sections. In addition to this, we will also discuss some practical experiences from safety and hazard analysis, namely, that the results will vary quite a lot, depending on who does the analysis—their background, their perspective, and the way they use a method.

We need to keep in mind the difference between a hazard and a risk. A hazard has the *potential to cause* death, injury, damage, or other loss. When we lose control, the hazard will cause an unwanted event. The risk level is the product of an event's likelihood and severity.

Since the bullet list above contains both methodology and method, a simple clarification might be in order. To put it simple terms, a method is a tool, while a methodology is a job description. More formally:

- *Method* is a tool, a component of research—say, for example, a qualitative *method* such as interviews.
- *Methodology* is the justification for using a particular *method*.

7.4.2 *What a Hazard and Risk Analysis Report Should Contain*

7.4.2.1 A Description of the System

There are many ways to describe a system—e.g., UML diagrams, state charts, or block diagrams. The choice will depend on the information available at the current stage of development. When we do the first safety analysis, we most likely only have a system sketch and a short description of what the system shall do.

Depending on the state of development, the safety analysis will have one of the following goals:

- Assign a SIL or ASIL level to the system in order to identify important development process activities.
- Identify the system's safety requirements.
- Check that the system is safe for the intended use and ready to be installed.

It is important to identify the system's operating environment and how the system interacts with its environment. In addition, we need to remember that a system description can have many forms. For example, see Fig. 7.8 where the left-hand and the right-hand diagrams describe the same system. In addition, both diagrams were used by a company to do safety analysis—see Stålhane and Malm (2020).

7.4.2.2 The System's Scope and Boundaries

The system's scope and boundaries give important pieces of information:

- What shall the system do—what is its functionality?
- What shall the system do it to and from—what is the system's input and operating environment?
- Are there operators involved, and if yes—what is their responsibility, and how much training and experience do they have? In addition—will other persons have access to the machinery?
- What is inside the system—the developers' responsibility—and what is outside—the customer's responsibility? For example, if the machine is dangerous, who is responsible for fencing it in?

7.4.2.3 Assumptions

It is both necessary and practical to make assumptions during a hazard analysis or safety analysis. It is, however, also important to document them. In addition, we need to document why we make these assumptions and what will happen if one or more of them have to be dropped. It is practical to consider assumptions and limitations—issue 10—together.

It is practical to state each assumption as "In order to be able to . . . we will assume that assumption X holds." Thus, we know what needs to be redone or reassessed if the assumption no longer holds.

7.4.2.4 Methodology Used for Safety and Hazard Analysis

We need to describe how we have attacked the problem at hand. This includes a lot of choices that have to be described. We also need to provide rationales for our choices. Thus, we need to describe, e.g., what level of details is used in the analysis and which parts of the system have we considered. The rationales for a certain set of guidewords (HAZOP) or failure modes (FMEA) are also part of the methodology.

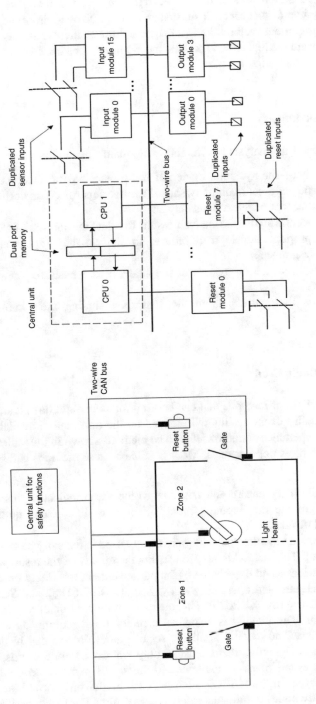

Fig. 7.8 Two ways to describe the SafeLoc system

If we use a fault tree approach, we have to decide if we will make one fault tree for the whole system or one for each subsystem and decide on appropriate subsystems.

Another important methodological choice is how to involve the users. Remember, the users are the real experts on how the system will be used and may have a lot of important information to share.

7.4.2.5   Stakeholders

The hazard and risk analysis has the following main stakeholders:

- The customer—the system's owner. If the project has no customer, someone inside the development company should assume this role. The customer should review and accept or reject the safety analysis.
- The developers who use the analysis results to select the appropriate SIL and thus also the appropriate development process. In addition, they implement barriers to make the system safe.
- The testers, who use the results of the analysis as a basis for creating safety-related tests.
- The experts that participate in the analysis including their knowledge and experience.

7.4.2.6   Methods Used

There is a plethora of methods that can be used for safety and hazard analysis. The choice and quality of the results will depend on the information available and the participants' experience and knowledge. The methods shown in the bullet list below are among the most popular ones. More details on these methods can be found in Chap. 6.

- PHA—Preliminary hazard analysis. Just as the name indicates, this analysis is used early in the development process. Even so, it contains quite a lot of important information.
- HazId—Hazard identification. They are mostly used in the early phases of system development. The method is easy to use, but the quality of the result will depend on the knowledge and experience of the participants in the HazId process.
- HAZOP—Hazard and operability study. See also IEC 61882 and SCSC's Data Safety Guidance (SCSC, 2020). This method can be used in all phases of system development. The method is heavy on ceremony and requires an experienced process leader. The method applies a set of guidewords to cue in the process participants. Originally, there existed only on set of guidewords, but with increased use and practical experiences, the set of permissible guidewords is now quite large. In the HAZOP analysis, each guideword is combined with the selected study nodes to initiate a safety consideration. For example, if we use the

guideword "No" and the study node "sensor output," we will get the issue "What happens when there is no sensor output?".

- FMEA—Failure Mode Effect and Analysis. This method has many things in common with HAZOP but is not so heavy on ceremony and uses failure modes instead of guidewords to help the participants to focus. There is a large set of available failure modes—from the simple pair "wrong response"-"no response" to the rather large set suggested by NCR—see NRC (2013). An example is shown in Table 7.4.
- FMEDA—Failure Mode Effect and Diagnostics Analysis also includes the failure detection probability and the detection methods used.
- IF-FMEA—Short for input-focused FMEA. IF-FMEA shows both possible failures stemming from the component itself and failures due to erroneous input to the component from sensors or from other system parts.
- FTA—Fault tree analysis is a simple, intuitive method that can be used to analyze safety and reliability—see IEC 61025. We will, however, focus on the safety part. The idea is simple—we start by asking "how can this system fail?". The answers will be documented in a tree fashion using "AND" and "OR" gates to show how events may be combined. We can then ask the same question for each of the component on the next level and so on. The result is an easy-to-interpret diagram.

Instead of analyzing a component, we can analyze a function. The table for functional FMEA looks just like the ordinary FMEA but applies the failure modes to a function instead of a component. Several of the methods mentioned above can also contain estimates of probability and consequence—either graded as "high, medium, or low" or with a score between 1 and 10. These two scores are used to compute a risk estimate (RPN—Risk Priority Number), which is used to prioritize the identified risks.

For more information on methods for safety analysis, see Chap. 6 "Safety Analysis Methods Applied to Software."

7.4.2.7   Data Sources

A safety or hazard analysis uses a lot of information and data. Some of the more important ones are:

- The currently available definition of the system that we will analyze
- The hazard log, which will give information on what has gone wrong in the past and how it was handled
- Generic failures, e.g., generic fault trees—general or domain specific

If we use reliability assessments—e.g., high or low in FTA—there must be a rationale for the choices. The same holds for the assessment of failure consequences and probabilities. For hardware components, there are many data sources, both from the producers and from several organizations—see, for instance, the PDS data handbook (Håbrekke et al. 2013) and IEC 62380:2004.

7.4.2.8   Guidewords, Failure Modes, Etc.

As mentioned for issue 6, we need to document the failure modes and/or guidewords used in the analyses. The most commonly used HAZOP guidewords are "no, more, less, as well as, part of, other than, and reverse." However, smaller sets such as "no, wrong" are also used in the industry.

There are two ways to look at failure modes: (1) the choice of failure modes depends on the unit we are analyzing, and (2) we have a set of generic failure modes that can be used for all units—hardware and software. We recommend the analysts to use a generic set of failure modes. Just to clear up a popular misconception—a failure mode is not an error/failure but how the error/failure manifests itself in the running system. Thus, "no response" is a failure mode, while the reason we do not get a response is an error/failure.

For more information on HAZOP and guidewords, see Chap. 6 "Safety Analysis Methods Applied to Software."

7.4.2.9   Summary of Results and Recommended Actions

The output from a hazard and risk analysis should be:

- Identified hazards and risk
- Actions that may be used to control or remove the hazards and risks identified
- Actions that may help the development company to improve its development process

7.4.2.10   Limitations of the Analysis

The limitations are important since they tell us where and when and where the risk or hazard analysis is valid. The limitations may be related to operational modes; physical environment, e.g., temperature, EMC, or vibrations; or software environment—e.g., libraries, operating system, or networks. When we are referring to environmental software components, it is important to include version ID. It is practical to consider limitations and assumptions—issue 3—together.

The limitations must be clearly stated so that the stakeholders know when and where the analysis can be trusted.

7.4.2.11   Analysis Records

The records that must be made available are:

- The analysis participants, including information about their training and experience.
- The results from the analysis—e.g., FMEA tables or HAZOP tables

- The recommendations related to safety—e.g., actions needed to avoid or control the identified hazards
- New information to be included in the company's hazard log

7.4.2.12 Any Other Information

Even though this section is not included in the referenced standard, we consider it important to include a set of definitions related to the terms used both in the analysis and in the presented results.

In addition to the requirements mentioned in IEC 60812:2018, it is also useful to consider ISO PAS 21448, which defines road vehicles, safety of the intended functionality, and this standard's discussions on SOTIF—safety of the intended functionality.

Table 7.7 Comments to some safety analysis standards

Topics	Experts/team	Plan	Process	Documentation
RCA IEC 62740:2015	Selection of a relevant team is an important part of the standard	A few requirements and some information are included	Information is included	Documentation is mentioned
HAZOP IEC 61882:2016	Selection of a relevant team is an important part of the standard	This is an important part of the standard. What the plan should include is also listed	A process for the HAZOP is well written, and figures are included	Documentation is presented in Sect. 6.5 and states that this is one of the strengths by using HAZOP
FTA IEC 61025:2006	Not emphasized	Some information is included	Some information is included	Requirements for the FTA report are presented in Chap. 9
FMEA IEC 60812:2018	Selection of relevant experts is an important part of the standard	This is an important part of the standard. What the plan should include is also listed	Some information is included	Requirements for the FMEA report are presented in section 5.4 in IEC 60812:2018
Design review IEC 61160:2005	Selections of relevant experts (specialists) are an important part of the standard. Top management shall be included	This is an important part of the standard	A process for the design review is well written, and a figure is included	Documentation is mentioned
RBD IEC 61078:2006	Not emphasized	Not emphasized	Not emphasized	Not emphasized

The table below shows some comments to standards relevant for hazard and risk analysis reports. In addition, the table shows the information on what the respective standards give us of help in choosing experts for the analysis, how the analysis should be planned, the needed process, and requirements related to the documentation (Table 7.7).

References

Friedrichsen, U.: Resilience reloaded. More resilience patterns. Codecentric AG (2015–2016)

Håbrekke, S., Stein Hauge, S., Onshus, T.: PDS Data Handbook 2013 Edition—Reliability Data for Safety Instrumented Systems (2013). isbn:9788253613345

Hanssen, G.K., Haugset, B., Stålhane, T., Myklebust, T., Kulbrandstad, I.: Quality Assurance in Scrum Applied to Safety Critical Software. XP 2016 Edinburgh (2016)

Hanssen, G. K., Stålhane, T., Myklebust, T.: SafeScrum—Agile Development of Safety-Critical Software. Springer, Cham (2018). isbn:9783319993348

IHASCO.: www.ihasco.co.uk/blog/entry/2506/what-is-a-dynamic-risk-assessment (2021)

Klim, H.Z., Balazinsky, M.: Methodology for Common Mode Analysis. SEA technical papers, 2007-01-3799 (2007)

Myklebust, T., Stålhane, T.: Safety Stories—A New Concept in Agile Development. SafeComp 2016-09, Trondheim (2016)

Myklebust, T., Eriksen, J.A., Hellandsvik, A., Hanssen, G.K.: The Agile FMEA Approach, SSS 18, York (2018)

NOG 070: Guidelines for the Application of IEC 61508 and IEC 61511 in the petroleum activities on the continental shelf (Recommended SIL requirements) Published: 15.04.2020 (2020)

NRC: Identification of Failure Modes in Digital safety Systems—Expert Clinic Findings—Part 2. Research Information Letter (2013)

Paige, R.F., Galloway, A., Charalambous, R., Ge, X.: High-integrity agile processes for the development of safety critical software. Int. J. Crit. Comput. Based Syst. 2(2) (2011)

Ritchey, T.: General Morphological Analysis—A General Method for Non-Quantified Modelling. Swedish Morphological Society (2002) (revised 2013)

RSSB GE/GN 8643: www.rssb.co.uk/standardscatalogue/CatalogueItem/GEGN8646-Iss-1 (2017). Last visited 27 April 2021

Schach, S.R., Jin, B., Yu, L., Heller, G.Z., Offutt, J.: Determining the distribution of maintenance categories: survey versus measurement. Empir. Softw. Eng. 8(4), 351–365 (2003)

SCSC: Data Safety Guidance. https://scsc.uk/SCSC-127E (2020)

Stålhane, T., Malm, T.: Four Perspectives on Safety Analysis. ESREL (2020)

Vinerbi, L., Puthuparambil, A.B.: Framework for Automation of Hazard Log Management on Large Critical Projects (2017)

Chapter 8
Software Documents

Go, go, go, said the bird: human kind
Cannot bear very much reality.

T.S. Eliot: Four Quartets

What This Chapter Is About
- Tools validation plan
- Tool process
- Release notes
- The change log
- Software architecture

8.1 Tools Validation Plan

8.1.1 Scope, Purpose, and Introduction

Software tools have become a more and more important part of software development. Both hardware and software engineers use development tools. A software tool can vary from a stand-alone software package to a suite of software tools integrated into a tool chain. The IEC 61508 series describes two types of software tools:

- Online tools, which run as part of the application and offline tools used during the development or manufacturing phases
- Online software tools, which have the same requirements as any other piece of software in the safety system

This chapter describes the validation of offline tools used to develop or test the software in the product or system. Offline software tools can be of many forms like test tools, linkers, compilers, code editors (e.g., visual studio code by Microsoft), GUI (graphical user interface), designer, assemblers, debugger, analysis tools, requirements management tools, etc.

© The Author(s), under exclusive license to Springer Nature Switzerland AG 2021   193
T. Myklebust, T. Stålhane, *Functional Safety and Proof of Compliance*,
https://doi.org/10.1007/978-3-030-86152-0_8

Tools must be validated when developing safety-critical products and systems. The importance and workload are related to which safety class the tool has (or similar requirement), depending on the relevant safety standard. It is important to consider the consistency and the complementary of the chosen tools, including the developers' understanding of and competence in the selected tools. This might be a challenge, for instance, for formal methods. The requirements according to the annexes for IEC 61508-3:2010 will depend on the SILx value.

8.1.2   Input Documents

In the Table 8.1, we have listed the relevant input documents and the related plans.

8.1.3   Requirements and Tool Process

8.1.3.1   Requirements in Safety Standards

In the last decade, we have seen a rapid development of relevant tools to be used together with the necessity of more tools as more and more companies move toward an agile and DevOps approach. The DevOps approach and Industry 4.0, e.g., OTA (over-the-air) tools and patching-related tools, result in the need to have more tools available.

The tool requirements in the ISO 26262:2018 series are more comprehensive than the current editions of IEC 61508:2010 and EN 50128:2011/A2:2020. The new editions of IEC 61508 and EN 50128 will include requirements similar to ISO 26262:2018 but probably also include more comprehensive requirements; see, e.g., Table 8.2, which is a draft of the next edition of IEC 61508-3. 1. Tools are

Table 8.1 Input documents and related plans

Input documents	Related plans

• Relevant safety standards, e.g., IEC 61508, ISO 26262, and EN 5012x series
  – Draft IEC 61508-3 ed.3
• Description of the system (DoS), including relevant SILx (ASILx) for the relevant parts and products
• Description of the ODD, including OEDR
• T&M; see Sect. 3.1
• Information from the tool supplier
• Tool version sheet
• User manual
• Information from the COTS supplier
	Part of this book
	  • Safety plan issued by suppliers; see Sect. 4.1
	  • SQAP: Software quality assurance plan; see Sect. 4.3
	Not part of this book
	  • Configuration management plan, IEEE Std. 828:2012 and ISO 10007:2017
	  • Certification plan (Myklebust 2013)
	  • Data safety management plan (SCSC 2021)

Table 8.2 Required tool confidence evidence per TIL

#	Required confidence data	TIL-0	TIL-1	TIL-2	TIL-3	TIL-4
1	Tool usage plan and classification	HR	HR	HR	HR	HR
2	Tool user documentation	R	HR	HR	HR	HR
3	Tool integration and validation in usage context	–	R	HR	HR	HR
4	Tool configuration management in usage context	–	R	HR	HR	HR

5a	Evidence of tool development for avoidance of systematic faults
	  – Architecture specification and verification
	  – Module design and testing with coverage
		–	–	R	HR	HR

5b	Evidence of confidence from tool usage history	–	R	HR	R	R
6	Tool problem management	–	–	R	HR	HR

relevant both when developing hardware and software. EN 50129:2018 chapter 6.3 includes requirements for safety-related tools for electronic systems.

8.1.3.2   Proven in Use

There are several definitions of the term "proven in use," and each safety standard seems to have its own. Below are some examples from standards commonly used for safety-critical software. Note that the definitions used in the Norwegian Oil and Gas industry standard use the two terms "proven in use" and "prior use" as separate criteria.

- IEC 61508:2010 Proven in use demonstration, based on an analysis of operational experience for a specific configuration of an element, that the likelihood of dangerous systematic faults is low enough so that every safety function that uses the element achieves its required safety integrity level. In addition, we should consider IEC 61508-3-1, which is dedicated to reuse of previously developed software.
- ISO 26262:2018 Proven in use argument—evidence that, based on analysis of field data resulting from use of a candidate, the probability of any failure of this candidate that could impair a safety goal of an item meets the requirements for the applicable ASIL.
- NOROG 070:2020—Norwegian oil and gas: The concepts of proven in use and prior use may look similar, but they have some important differences.

  - Proven in use is a concept introduced in IEC 61508-2 as an alternative route to demonstrate avoidance and control of systematic failures and applies to manufacturers of devices.
  - Prior use is a concept introduced in IEC 61511-1 for end users to qualify devices not developed according to IEC 61508.

- IEC 61511:2016 Prior use: documented assessment by a user that a device is suitable for use in a SIS (safety instrumented system) and can meet the required functional and safety integrity requirements, based on previous operating experience in similar operating environments.
- IEC 62304:2015 SOUPs (Software Of Unknown Provenance)—software that is already developed and generally available and that has not been developed for the purpose of being incorporated into the medical device or software previously developed for which adequate records of the development.
- EN 50657:2017 Pre-existing software: all software developed prior to the application currently in question, including commercial off-the-shelf (COTS) software, open-source software, and software previously developed but not in accordance with this European standard.
- ED-12C/ED-109A:2012 uses the term "previously developed software." PDS encompasses any software developed for use on another application. This includes commercial off-the-shelf (COTS) software products and software developed to previous or current software safety standards.

In addition to IEC 61508, we have looked at the handling of PIU (proven in use) for the standards ISO 26262, IEC 61511, IEC 62304, EN 50657, and ED-12C/ED-109A. Most of these standards refer back to IEC 61508. The automotive standard—ISO 26262—states this quite clearly: "The ISO 26262 series of standards is the adaptation of IEC 61508 series of standards to address the sector-specific needs of electrical and/or electronic (E/E) systems within road vehicles."

When deciding to use the PIU argument, the following issues should be considered for all the reference standards and leave documented information:

- Complete documentation of the component—e.g., functionality, space needed, and operational conditions.
- The safety requirements of the PIU component compared to the safety requirements in the system where the PIU component shall be reused. We should also do an analysis to see how a failure in the PUI component will affect the overall system's safety. An important input here is the data for all previous PIU component failures, how they have been analyzed, and the analysis results.
- The environment where the PIU component was previously used—preferably including hardware, software, and wetware (human aspects). These aspects must be compared to the environment where it will be used next.
- Assessment of PIU components suitable for use. This assessment must be available as documented information.

In addition to the usual "proven in use" components, ISO 26262 introduces the term "safety elements out of context" (SEooC)—see ISO 26262-10, clause 9. A SEooC is a safety-related element that is not developed for a specific item—i.e., it is not developed in the context of a particular vehicle. It must, however, be developed according to the ISO 26262 standard. As an example, consider a generic wiper system with assumed safety requirements to be integrated in different OEM (original equipment manufacturer) systems.

The ASIL capability of a SEooC designates the capability of the SEooC to comply with assumed safety requirements assigned with a given ASIL. Consequently, it defines the requirements of the ISO 26262 series of standards that are applied for the development of this SEooC.

The ISO 26262 is special, compared to the other standards cited here, with a detailed description of how to assess the safety and reliability of a PIU component—see the section on "Number of Hours of Operation." Before doing any reuse of automotive software, the developers should check ISO 26262, parts 8 and 10.

IEC 61511 and IEC 62304 only discuss modifications when it comes to how to avoid modifications of a PIU component. ISO 26262-8, however, discusses how modifications of PIU components shall be handled. Any modification to a proven in use item or element shall comply with ISO 26262-8, clause 14.4.4, for the corresponding proven in use credit to be maintained. A description of the candidate and its previous use must first and foremost contain an identification of the candidate with a catalogue of internal elements or components. Next, we need the functional requirements that describe the interface and environmental, physical and dimensional, and functional and performance characteristics of the candidate. We also need the safety requirements of the candidate in the previous use and the corresponding ASILs.

Suppose we modify an item or its environment. In that case, we need an impact analysis of the item to identify and describe the modifications applied to the item, including modifications to the design, implementation, or environment. The impact analysis shall evaluate the implications of the modifications with regard to functional safety and identify and describe the safety activities to be performed based on the impact of the modifications. If we do modifications to a candidate introduced after its evaluation period, we need to provide evidence that the proven in use status remains valid.

The ED-12C/ED-109A also provides a detailed discussion of modification. The first assumption is that the PDS has been previously incorporated into a certified system. Therefore, the process used to change the PDS baseline should be the same process as used when changing any other software product that has to comply with ED-12C/ED-109A. This includes evaluating the impact of PDS requirements changes on system requirements, the system safety analysis, and the previously accepted certification. In addition, we need to analyze the impact on software lifecycle data.

Only ISO 26262 and Norwegian Oil and Gas (NOROG) 070 uses the number of operating hours as a means to calculate reliability. The ED-12C/ED-109A only mentions that the number of hours of operation is relevant, while IEC 61508 only requires a failure rate better than 10^{-5}/h. However, IEC 61508-7 has a set of requirements for field experience data. These requirements are also used by Norwegian Oil and Gas 070 and are as follows: For field experience to apply, the following requirements must be fulfilled:

- Unchanged specification
- Data from ten systems in different applications

Table 8.3 Tool validation—proven in use

ASIL	Observable incident rate	Minimum observation period without observable incident	Observable incident rate Interim period
D	$<10^{-9}$/h	1.2×10^9 h	$<3 \times 10^{-9}$/h
C	$<10^{-8}$/h	1.2×10^8 h	$<3 \times 10^{-8}$/h
B	$<10^{-8}$/h	1.2×10^8 h	$<3 \times 10^{-8}$/h
A	$<10^{-7}$/h	1.2×10^7 h	$<3 \times 10^{-7}$/h

- 10^5 operating hours
- At least 1 year of service history

ISO 26262 uses the table and formula below. Note that the table is a combination of Tables 6, 7, and 8 in ISO 26262-8:2018 and that the standard has no differences between ASIL B and C when it comes to incident rates (Table 8.3).

The table above is used as follows: The evaluation period shall demonstrate compliance with each safety goal that can be violated by the candidate in accordance with part 1 of the table—Observable incident rate. The column marked "Minimum observation period without observable incident" is an example.

$$t_{\text{service}} = \widehat{t_{\text{MTTF}}} \left(\frac{X^2_{0.7,df}}{2} \right)^2, df = 2f + 2,$$

where X^2 denotes the chi-square distribution and f is the number of safety-related incidents.

An alternative is to use the evaluation period defined in "Observable incident rate—Interim period." In this case, the candidate shall demonstrate compliance with each safety goal that can be violated by the candidate in accordance with the last column with a single-sided lower confidence level of 70% using a chi-square distribution.

In all cases, 1.2×10^9 h for ASIL D is a long time to test anything when we consider that a year that is not a leap year is a little less than 9000 h—8760 to be exact. The way out of this is to use equipment-years which is "number of units" $\times$ "number of hours in use" if no failures have occurred. Consider the following three examples:

- 100 valves

  - ASIL D: 1.3×1000 years—not practical
  - ASIL A: 13 years—not practical but at least remotely possible, e.g., as part of a service agreement

- 20,000 detectors

  - ASIL D: 65 years—not practical
  - ASIL B: 6.5 years—not practical but at least remotely possible
  - ASIL A: 7.8 months—definitely doable

- 100,000 detectors (e.g., relevant for the automotive industry)

  - ASIL D: 13 years
  - ASIL C: 1.3 years—not practical but at least remotely possible
  - ASIL B: 1.3 years—not practical but at least remotely possible
  - ASIL A: 1.6 months—definitely doable

It might be possible to claim that "We do not need PIU since the component is certified" or "We do not need certification since the component is already proven in use." ED-12C/ED-109A has as an assumption that before allowing the modification of a PDS, it must have been previously incorporated into a certified system. However, according to W. Goble (2017), these two methods for creating trust are not competing methods. The certificate shows that the component has been developed according to a specified standard and that a certification authority has validated this.

The problem is that the number of possible paths through a software system and thus the number of test cases needed is large—much too large to be completely tested. To quote R. Awedikian (2009): "Testing exhaustively a software product is a NP-Complete problem from a computational viewpoint. In other words, it is complex to test all the inputs, combinations of inputs and paths of software." A problem is NP-complete when a brute-force search algorithm can solve it and the correctness of other problem with similar solvability. The name "NP-complete" is short for "nondeterministic polynomial-time complete" (Wikipedia).

A much simpler argument runs as follows: Let the failure probability be p, the number of tests be n, and α be the significance level. Then we have

$$(1 - p)^n = \alpha \text{ and thus } n \approx |\frac{\ln \alpha}{p}|$$

If we, e.g., want $p < 10^{-6}$ and a significance level of $\alpha = 0.05$, we get $n \approx 3 \times 10^6$ independent tests, which might be close to unachievable for most practical test campaigns. One way to increase confidence in a component is that it has been used in a similar environment earlier. This will not show that the component is fault-free, but it will show that a lot of usage-dependent—and thus environment-dependent—failures have been discovered and corrected. This is the reason why all relevant standards in one way or another specify that a PIU component has earlier been used in a similar environment.

The certificate will show that the component is developed with a sound architecture and design and that it is developed and tested by competent personnel. However, only the "proven in use" argument will show that it is error-free—at least for the planned environment.

As a general set of rules, in order to use the "proven in use" argument for a software component, you need to have:

- Documentation of the component and preferably the process that was used to develop it
- The component's safety requirements
- A description of the environment(s) where it has previously been used and

  - The duration of its use in this environment
  - The number of safety-critical incidents during this period

- Assessment of the component's suitability for the new (intended) environment

As shown in the previous sections, the details for claiming PIU will vary from standard to standard. If you want to change a PIU component, it is important that you do a change impact analysis and make sure that previously made assumptions still are valid. As always—check with the assessor.

8.2   Tool Process

The first part of a project's tool process identifies which tools can be beneficial for the project.

This work will identify the relevant tools based on, among other things:

- Available tools within the company
- Company tool strategy
- Relevant SIL/ASIL for the product or system
- T&M to be applied

  - Safety analysis tools
  - Static code tools, etc.

- The process to be applied, e.g., waterfall, V-model, or DevOps
- Relevant tool chains
- Security aspects

The process for the different safety standards is presented in the Fig. 8.1.

Many tools are relevant. Some can be related to the V-model approach. Below, we have listed relevant tools using a V-model approach (Fig. 8.2).

If having a DevOps process, other tools have to be used. In the figures below, some of the top DevOps tools currently available in the market are presented (Figs. 8.3 and 8.4).

DevOps practitioners have developed a "periodic table" for relevant software tools. More than 1800 practitioners cast their votes for 400 products in 17 categories to produce the 2020 periodic table shown below. See also https://digital.ai/

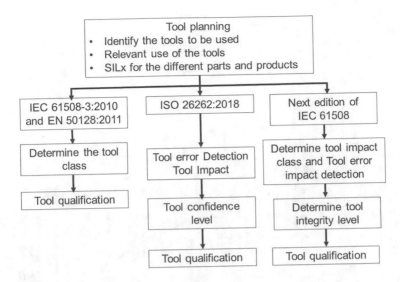

Fig. 8.1 Tool qualification process

periodic-table-of-devops-tools. The colors show different tool groups. For example, dark blue is for security. In Fig. 8.5, we have explained the information for each tool. In addition, when using the Internet page, one may click on each toolbox to receive more information regarding each tool and similar tools.

The second phase is to determine the usage of the tools; this includes:

- How the tool will be used. This depends on the tool's functions and properties.
- Adaptation of the tool. The appropriate adaptation depends on the project and context. This may be related to a tool chain and, e.g., a link between a requirement tool and a UML use case and also a link to the relevant hazard log ID.
- Integration of the tool into the user environment and as part of, e.g., a tool chain.
- Tool limitations. Limitations could be related to the analysis capability, test capability, and, e.g., modelling capabilities.
- Appropriate use of the tool according to the tool user manual and relevant context and project information.

The next phase is to determine the relevant tool classes (IEC 61508:2010 and EN 50128:2011) or tool confidence level (ISO 26262).

Examples of tool type and the related tool classification are presented in the Table 8.4:

The third phase is to qualify the tools.

There are different methods to qualify a tool. The method is related to the SIL and the environment in which the software tool is executed. All offline support tools in classes T2 and T3 shall have a specification or product documentation that clearly defines the tool's behavior and any instructions or constraints on its use. Validation measures shall provide evidence that the software tool complies with specified

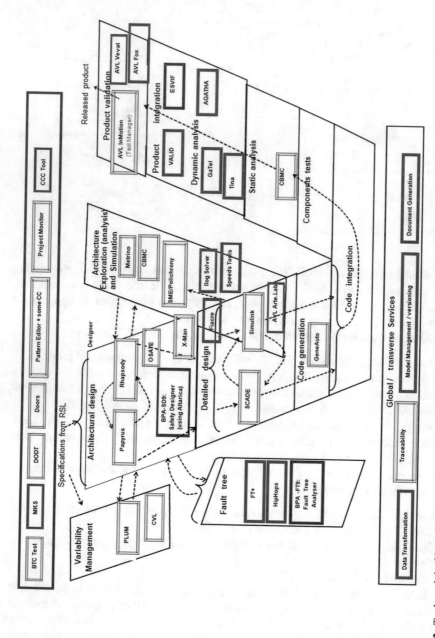

Fig. 8.2 Tools and the V-model

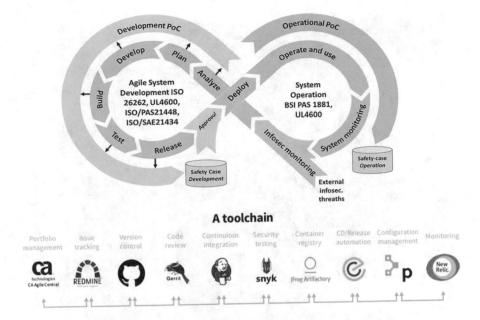

Fig. 8.3 DevOps and related tools

requirements to its purpose. The appropriateness of tool user training and competency requirements shall be considered too.

Relevant methods are:

- Certified tools.
- Proven in use (confidence from use). Used for the same purpose and comparable use cases:

  - It has been confirmed by testing and periodically checked by analysis including relevant evidence that the tool is suitable for the purpose for which it is to be used.
  - There is sufficient relevant evidence from previous use to show that the output of the tool can be trusted, including a suitable combination of documented history of successful use (project and applications realized, list of bugs, etc., identification of successive versions).

- Evaluation of the tool development process.
- Version history may provide assurance of maturity of the tool and a record of, e.g., the errors and ambiguities that should be taken into account when the tool is used in the new development environment.
- Software HazOp.
- Data HazOp if the tool includes data-intensive elements.
- Compatibility of the tools of an integrated toolset.
- Configuration management. Each new version has to be qualified.

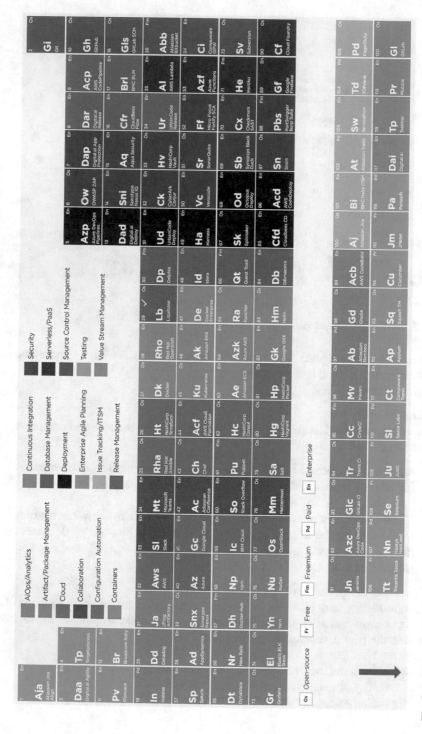

Fig. 8.4 The periodic table of relevant DevOps tools. ITSM: information technology service management. PaaS: platform as a service. © Digital.ai

Fig. 8.5 Explanation for
the Red Hat Ansible toolbox

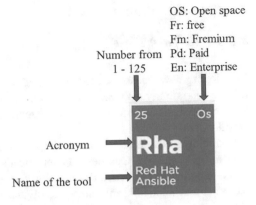

OS: Open space
Fr: free
Fm: Fremium
Number from   Pd: Paid
1 - 125   En: Enterprise

Acronym

Name of the tool

25   Os
Rha
Red Hat
Ansible

Table 8.4 Tool type and the related tool classes

Tool type	Classification	Comments
Scrum workflow management	T1	Relevant tool is Jira
UML modelling	Tx	Depends on the use T1 if used for illustration T2 if used for verification T3 if code is automatically generated
Design documentation	T1	Relevant tools are Doxygen, Word, Excel
Requirements/test management	T1 T2	Relevant tools are Rmsis, Doors, Excel T2 if used for, e.g., testing or verification
Static code analysis	T2	Relevant tools are QAC/QA-C++

Relevant mitigation measures are:

- Avoiding known bugs and errors
- Restricting use of the tool functionality
- Checking the tool output
- Using diverse tools for the same purpose

The last part of the validation is to document the validation. Relevant topics are:

- Record of the test, verification, analysis, and validation activities.
- The version of the tool product manual being used.
- The tool functions being validated.
- Tools and equipment used.
- The documented results of validation shall state either that the software has passed the validation or the reasons for its failure.
- Discrepancies between expected and actual results.

8.3   Release Plan and Notes

8.3.1   Release Plan

Developing new releases of safety-critical software is more wide-ranging than for release of ordinary software as it requires, among other things, impact analysis, testing, analysis, configuration management, validation, and in many cases certification.

Planning software releases is an important part of the overall planning process. Due to agile and DevOps development needs, we will split the releases into two parts—internal and external. Before a software release, the software baseline shall be recorded and kept traceable under configuration management control so that it is possible to reproduce each software release—both external and internal:

- Internal releases are aimed at developers for testing and analysis. Software is integrated and released for testing as soon as it is uploaded to the integration servers. We should re-run previous integration tests, FATs, and SATs and new tests for the change.
- External releases are meant for the customers and may only be released after proper testing, analysis, and certification of the safety-critical functions. There are two types of external releases:

  - Major user releases based on a stabilized development
  - Minor releases used to address minor bugs, security issues, or critical defects

A release plan with fixed dates upfront will promise the customer(s) to deliver a certain set of functionalities at a specific time. This can be achieved by controlling the story priorities—whatever shall be in the next release must have the highest priorities.

Consequently, if one of them is changed, the developers must coordinate the release plan and the story priorities (including safety stories).

Regression testing must be included in the plan for each release. Regression testing is needed to show that (1) the latest changes did not introduce an error in already existing functionality and (2) the changes did not re-introduce already fixed errors.

External releases shall come with a release note which shall include information on:

- All restrictions in using the software. Such restrictions are derived from, for example, non-compliances with standards or lack of fulfilment of all requirements.
- The application conditions.
- Compatibility among software components and between software and hardware.

8.3.2 *Release Notes*

The release note tells the user what is new, the fixed problems, and how the system shall be installed and maintained. The release notes should be written in a clear language and be kept to the point.

Release notes were originally written for a situation where a new version of the software was released two to four times a year. DevOps has an explicit goal to release a new version of the software as quickly as possible after a problem has been reported from an operator and fixed. This will put new requirements on the release note process when it comes to speed. If the system is safety-critical and independent safety assessment is needed—e.g., for railway—there is a real challenge. More research and cooperation with certification authorities are needed to handle quick releases, release notes, and independent safety assessments quickly and efficiently.

As standards and guidelines, we have used the following documents:

- IEEE 828:2012 Standard for Configuration Management in Systems and Software Engineering
- EN 50128:2011 Railway applications—Communication, signalling, and processing systems—Software for railway control and protection systems
- ProductPlan: How to Write Release Notes Your Users Will Actually Read (ProductPlan 2021)
- IEC 61508-3:2010 The techniques and measures
- ISO 26262-2:2018 Road vehicles—Functional safety—Management of functional safety

For IEEE 828, the main input to the release notes is the build manifest which identifies the contents of the build—i.e., metadata that describes the assembly itself—name, version, required external assemblies, etc. In addition, we will need the requirements—hardware and software—and the test log for the new version.

EN 50128 states that the release notes should be written under the responsibility of the designer, on the basis of all design, development, and analysis documents relevant to the deployment. In addition, the release notes shall go through a verification process to establish that the references in the release notes are unique and that correct acronyms and abbreviations are used.

ISO 26262-2 states that the documentation of functional safety for release for production shall include the following information: (1) the name and signature of the person responsible for the release, (2) the versions of the released item or elements, (3) the configuration of the released item or elements, and (4) the release date.

EN 50128 has much more strict requirements for the release notes than IEEE 828 has. The main reason for this is probably that IEEE is not concerned with certification. According to IEEE 828, the release note should contain the requirements for the installation environment, installation instructions, and a summary of changes since previous release. According to EN 50128, the release note should contain an overview of all restrictions in using the software, information on compatibility, and a description of the application conditions. ISO 26262-2 adds that the release note should contain, or have reference to, a safety case.

We suggest the following contents for a release note:

- Release version number
- Requirements related to hardware and other software—e.g., the operating system
- A statement on the new release's compliance with relevant standards
- References to installation and deployment instructions
- Changes in the system's functionality—what it can do now that it could not do before and what it cannot do anymore
- Changes related to safety and security
- An updated safety case
- Changes to the application conditions (AC) and the safety-related application conditions (SRAC)
- Reported and corrected problems (e.g., errors and bugs)
- Reported but remaining problems

ProductPlan has published some good advices to be used when writing a release note: Use plain language and keep them short. Group the items logically and include relevant links.

This process has two outputs—the release note and the release note assessment. The results of the release note assessment should be part of the software validation report.

8.4 Change Log

We first need to consider the fact that none of the standards discussed here—e.g., ISO 9001, ISO 26262, or any IEEE standard—use the term "change log." The closest we get is "Control of documents." Examples are ISO 9001, 8.5.6 Control of changes; ED-12C, 7.2.4 Change control; or ISO 26262-8, 8.4.5 Implementing and documenting the change.

- ISO 9001: The organization shall retain documented information describing the *results of the review of changes*, the person(s) authorizing the change, and any necessary actions arising from the review.
- ISO 26262: The documentation of the change shall contain the following information:

  (a) *The list of changed work products, items, and elements at an appropriate level including configurations and versions*
  (b) *The details of the change carried out*
  (c) The planned date for the deployment of the change

- ED-12C: Change control activities include:

  (a) Change control should preserve the integrity of the configuration items and baselines by providing protection against their change.
  (b) Change control should ensure that any change to a configuration item requires a change to its configuration identification.
  (c) *Changes to baselines and to configuration items under change control to produce derivative baselines should be recorded, approved, and tracked. Problem reporting is related to change control, since resolution of a reported problem may result in changes to configuration items or baselines.*
  (d) Software changes should be traced to their origin and the software lifecycle processes repeated from the point at which the change affects their outputs. For example, an error discovered at hardware/software integration, which is shown to result from an incorrect design, should result in design correction, code correction, and repetition of the associated integral process activities.
  (e) *Throughout the change activity, software lifecycle data affected by the change should be updated, and records should be maintained for the change control activity. The change control activity is aided by the change review activity.*

The change log entry must contain a reference to the relevant change impact analysis or a document showing the decision that a change impact analysis was not needed. In addition, we need a reference to the version number of the system plus the computer configuration on which it was run. The exact configuration is needed both to reproduce the problem and to check that it has been fixed.

Since there is no standard for a change log, we have taken ideas from the three standards above and suggest the following contents:

- The reason(s) for the change
- The person(s) authorizing the change
- The person(s) who performed the changes
- The details of the change carried out—e.g., changed work products, items, and elements at an appropriate level, including configurations and versions
- Reference to tests run to validate the changes
- Changes to baselines and to configuration items under change control needed to produce derivative baselines
- The person(s) who reviewed the changes
- The results of the review of changes and any necessary actions arising from the review
- Software lifecycle data affected by the change—if any

8.5   Architecture Documents

8.5.1   Software Architecture Specification

The software architecture describes how the system is organized. According to IEC 61508-3 "The software architecture defines the major elements and subsystems of the software, how they are interconnected, and how the required attributes, particularly safety integrity, will be achieved. It also defines the overall behaviour of the software, and how software elements interface and interact."

Two things follow from this: (1) the software architecture must be decided early in the development process and (2) it is extremely difficult/costly to change it later in the process.

- IEEE 42010: 2011: Systems and software engineering—Architecture description
- IEC 61508-3:2010: Functional safety of electrical/electronic/programmable electronic safety-related systems—Part 3: Software requirements

Table A2—Software design and development—software architecture design—in IEC 61508-3: 2010 lists a set of issues that have to be considered when defining the system's architecture. We have only listed issues that are highly recommended (HR) for SIL 3.

- Fault detection and graceful degradation
- Modular approach
- How safety and non-safety parts of the system are separated
- Forward and backward traceability between the software safety requirements specification and software architecture
- Structured diagrammatic methods or semi-formal methods (e.g., UML)
- Computer-aided specification and design tools
- Cyclic behavior, with guaranteed maximum cycle time, time-triggered architecture, or event-driven, with guaranteed maximum response time
- Static resource allocation

Some of the items on the list are related to what to do—e.g., forward and backward traceability and how safety and non-safety parts are separated—while other items are related to how to do, e.g., fault detection and graceful degradations that can be achieved through redundancy and diversity.

The architectural description must contain architecture views and viewpoints, architecture models and relations, and architecture rationales and decisions. The architectural models and relations are made based on the IEC 61508-3 requirements in the bullet list above. There should be one rational for each architectural viewpoint where we explain the reasons for each architectural choice or decision.

Note the difference between view and viewpoint:

- View: A representation of a whole system from the perspective of a related set of concerns

- Viewpoint: A specification of the conventions for constructing and using a view. A pattern or template from which to develop individual views by establishing the purposes and audience for a view and the techniques for its creation and analysis

The two characteristics, coupling and cohesion, are also import when we want to assess a system's architecture. We will use definitions from IEEE 24765: 2010: "Cohesion, also known as module strength, may mean the manner and degree to which the tasks performed by a single software module are related to one another or—in software design—a measure of the strength of association of the elements within a module." Low cohesion would mean that the code that makes up some functionality is spread out all over your codebase.

Coupling has several meanings:

- The manner and degree of interdependence between software modules.
- The strength of the relationships between modules. ISO/IEC TR 19759:2005, Software.
- A measure of how closely connected two routines or modules are. In software design, coupling is a measure of the interdependence among modules in a computer program—SWEBOK 1.4.2.3 (SWEBOK 2015).

High coupling would mean that your module knows the way too much about the inner workings of other modules.

GeeksforGeeks (2021) has, however, two shorter definitions, which also are easier to use and understand:

- "Coupling is the measure of the degree of interdependence between the modules." Good software should have low coupling.
- Cohesion measures the degree to which the elements of the module are functionally related.

As we see from the diagram in Fig. 8.6, the ideal component has low coupling and high cohesion. We also see that cohesion is more important than coupling—a component can be good if the cohesion is low even when the coupling is high.

A simple way to check if there is too much coupling is to ask one simple question: If I want to make a change to component A, how many other components might I have to change? The larger the number you get as an answer, the worse is the coupling. A similar question may be used to check for cohesion: If I want to change function F, how many components might I have to change? The larger the number you get as an answer, the worse is the cohesion. Thus, in the end, coupling and cohesion are all about maintainability. ISO 25010 defines maintainability as follows: "The effectiveness and efficiency with which a system can be modified to improve it, correct it or adapt it to changes in environment, and in requirements." Maintainability has the following sub-characteristics:

1. *Modularity.* The system is composed of discrete components such that a change to one component has minimal impact on other components.
2. *Reusability.* An asset can be used in more than one system or in building other assets.

Fig. 8.6 Types of code
from cohesion and coupling
perspective

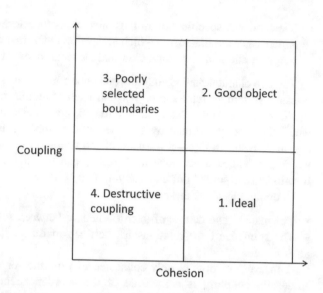

3. *Analyzability.* The effectiveness and efficiency with which it is possible to assess the impact on a system of an intended change to one or more of its parts, or to diagnose a product for deficiencies or causes of failures, or to identify parts to be modified.
4. *Modifiability.* A system can be effectively and efficiently modified without introducing defects or degrading existing product quality.
5. *Testability.* The effectiveness and efficiency with which test criteria can be established for a system and tests can be performed to determine whether those criteria have been met.

We see that all of the characteristics in the bullet list above are strongly related to coupling and cohesion. Thus, when inspecting an architecture to see if it is developed according to the process defined in the standard, it is not enough. We also must check for coupling—weak—and cohesion, strong. The requirements for coupling and cohesion have a lot in common with the architectural requirements for code adapted to DevOps. In addition to the five sub-characteristics required for maintainability, DevOps also requires:

1. Deployability
2. Monitorability
3. Configurability/configuration
4. Certifiable for safety and security

The output from this process is the architectural description, architecture views and viewpoints, architecture models and relations, and architecture rationales and decisions. The software architecture report is an important input for the software architecture and design report.

8.5.2   *Application Architecture and Design*

Wikipedia defines an application architecture as a description of "the behaviour of applications used in a business, focused on how they interact with each other and with users. It is focused on the data consumed and produced by applications rather than their internal structure."

In addition, an "applications architecture tries to ensure the suite of applications being used by an organization to create the composite architecture is scalable, reliable, available and manageable." Note that an application architecture "is different from software architecture, which deals with technical designs of how a system is built."

Most of this section is based on "The Open Group Architecture Framework" (TOGAF) (White 2018)—an enterprise architecture methodology that offers a high-level framework for enterprise software development. The reason why we chose TOGAF is that "TOGAF is perhaps the most popular Enterprise Architecture (EA) framework today, and its popularity is only increasing" (Bloomberg 2014). According to Open Group, "80 per cent of Global 50 companies and 60 per cent of Fortune 500 companies used the framework" (White 2018).

Note that even if this section is based on TOGAF, the relevant ISO standards still apply. ISO/IEC 42010 defines "architecture" as "The fundamental organization of a system, embodied in its components, their relationships to each other and the environment, and the principles governing its design and evolution." TOGAF embraces but does not strictly adhere to ISO/IEC 42010 terminology. In TOGAF, "architecture" has two meanings depending upon the context.

- A formal description of a system or a detailed plan of the system at component level to guide its implementation.
- The structure of components, their inter-relationships, and the principles and guidelines governing their design and evolution over time. It is this alternative that is of interest in this section.
- ISO/IEC/IEEE 42030: Software, systems and enterprise—Architecture evaluation framework.
- ISO 15704: Industrial automation systems—Requirements for enterprise-reference architecture and methodologies.
- What is TOGAF? An enterprise architecture methodology for business.

According to TOGAF, an enterprise architecture has four architecture domains: business architecture, data architecture, application architecture, and technology architecture.

According to ISO 15704:2000, enterprise-reference architectures and methodologies that are model-based shall include these four model-content views: function, information, resource, and organization.

The application architecture process will produce outputs such as process flows, architectural requirements, project plans, and project compliance assessments. The TOGAF Architecture Content Framework provides a structural model for architectural content. This allows the work products to be consistently defined, structured, and presented. The framework uses three categories to describe the type of architectural work product within the context of use:

- Deliverable—a work product that is contractually specified and in turn formally reviewed, agreed, and signed off by the stakeholders.
- Artifact—a more granular architectural work product that describes an architecture from a specific viewpoint—see Sect. 4.3.
- Building block—a component of business, IT, or architectural capability that can be combined with other building blocks to deliver architectures and solutions.

Two of the key elements of any enterprise architecture framework are a definition of the deliverables that the architecting activity should produce and a description of the method by which this should be done.

A TOGAF document model structures the release management of the TOGAF specification. It is not intended to serve as an implementation guide for practitioners. Within the model, the content of the TOGAF document is categorized as follows:

- Core—the fundamental concepts that form the essence of TOGAF.
- Mandated—the normative parts of the TOGAF specification. These elements of TOGAF are central to its usage, and without them, the framework would not be recognizably TOGAF. Strong consideration must be given to these elements when applying TOGAF.
- Recommended—a pool of resources that are specifically referenced in TOGAF as ways in which the TOGAF Core and Mandated processes can be accomplished (e.g., the SEI Architecture Trade-Off Analysis Method or business scenarios).

ISO/IEC/IEEE 42030: 2019, Annex C.5, shows an example of an enterprise architecture evaluation.

The application architecture and design need to be coordinated with the software architecture and design verification report.

8.5.3   Architecture Standard: An Example

How to make hardware and software cooperate is, among other things, an architectural question. The Fig. 8.7 shows one way to solve this by having an extra layer between customer software and OEM hardware. Thus, instead of having to adapt to each type of hardware, you only need to adhere to the AUTOSAR (Automotive Open System Architecture) standard, which will then take care of the rest—www.autosar.org/. The diagram shown below, taken from Das (2014), shows the main idea.

An articulated goal of the AUTOSAR consortium is to "Cooperate on Standards, Compete on Implementation" (Das 2014). AUTOSAR provides the following safety mechanisms:

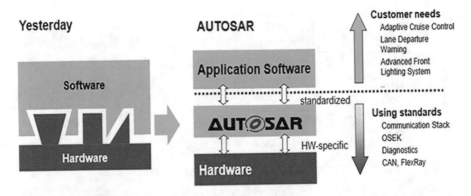

Fig. 8.7 AUTOSAR—the principal view, © Abhash Das

- Built-in self-test mechanisms for detecting hardware faults
- Run-time mechanisms for:
  - Detecting software faults during the execution of software—program flow monitor
  - Preventing fault interference—memory partitioning for SW-C (software components) and time partitioning for applications
  - Protecting the communication—end-to-end (E2E) communication protection for SW-Cs
  - Error handling

When using AUTOSAR, the safety requirements may stem from four sources:

- The ISO standard for automotive software—ISO 26262
- WGs (AUTOSAR Working Group) and WPs (AUTOSAR Work Package)
- Application requirements
- Safety requirements derived from the safety analysis

These four sets of requirements are combined into a consolidated set of safety requirements. Since we will combine ISO 26262 and AUTOSAR, we need to consider the rules on ASIL inheritance defined in ISO 26262. The AUTOSAR basic software and RTE inherit safety relevance either by implementing complete AUTOSAR basic software according to max ASIL of application software or demonstrating freedom from interference by software partitioning in basic software by appropriate mechanisms. When they have been consolidated, the safety requirements are again split up according to how they will be satisfied. They can be satisfied using:

- Process safety requirements—AUTOSAR
- Technical safety requirements
- Methodology safety requirements—tools and generation (Fig. 8.8)

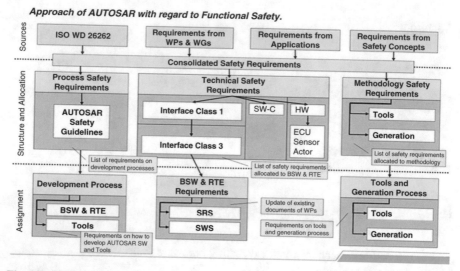

Fig. 8.8 AUTOSAR and functional safety, © Abhash Das

Both in the automotive industry and in many other application domains, we may use components that were not developed for the present application area and are developed by different organizations. As a consequence of this, the developers have to make assumptions on functional requirements and safety requirements. Such elements are called safety elements out of context—SEooC. Examples of SEooCs are a system controller, ECUs, micro-controllers, software implementing a communication protocol, or an AUTOSAR software component. These components are intended to be used when the validity of its assumptions can be established during integration of the SEooC. The components are not necessarily designed for reusability nor developed under ISO 26262.

References

Awedikian, R.: Quality of the Design of Test Cases for Automotive Software: Design Platform and Testing Process. Business Administration. Ecole Centrale Paris (2009)

Bloomberg, J.: Enterprise Architecture: Don't Be a Fool with a Tool. www.forbes.com/sites/jasonbloomberg/2014/08/07/enterprise-architecture-dont-be-a-fool-with-a-tool/?sh=5256ef717860 (Aug 2014)

Das, A.: Autosar. http://abhashr.blogspot.com/2014/03/autosar-approach-for-functional-safety.html (26 Mar 2014)

GeeksforGeeks.: www.geeksforgeeks.org/ (2021). Last visited 26 Apr 2021

Goble, W.: Tales from the Certification Wars—Proven In Use versus Certification CFSEW (19 Apr 2017)

Myklebust, T.: Certification of Safety Products in Compliance with Directives Using the CoVeR and the CER Methods. ISSC, Boston, MA (2013)

ProductPlan.: www.productplan.com/learn/release-notes-best-practices/ (2021). Last visited 7 June 2021

SCSC 127E: Safety-Critical Systems Club, Data Safety Guidance. Version 3.3 (2021)

SWEBOK.: http://swebokwiki.org/Main_Page (2015). Last updated, last updated 27 July 2015

White, S.K.: What Is TOGAF? An Enterprise Architecture Methodology for Business. https://www.cio.com/article/3251707/what-is-togaf-an-enterprise-architecture-methodology-for-business.html (2018)

Chapter 9
Test, Analysis, and V&V

A Turkey is fed for a thousand days by a butcher; every day confirms to its staff of analysts that butchers love turkeys with increased statistical confidence.

Nassim Nicholas Taleb—Antifragile: Things That Gain from Disorder

What This Chapter Is About
- Test reports
- Test specification and test scripts
- Software/hardware integration
- Software quality assurance verification report
- Architecture and design verification reports
- Software requirements verification reports
- Overall software test reports
- Software validation reports

9.1 FAT/SAT/CAT Report

9.1.1 Introduction

The alphabet soup used in the chapter heading stands for:

- FAT: Factory Acceptance Test.
- SAT: Site Acceptance Test. In maritime industry—Sea Trial Acceptance Test.
- CAT: Customer Acceptance Test. In some cases, the interpretation "Compliance Acceptance Test" is also used.
- SIT: Site Integration Test.

We will treat each of these test topics in a separate subsection below. First, we should define the terms we will use. Acceptance testing is about two issues—acceptance criteria and acceptance test process. According to IEEE 24765:2010,

© The Author(s), under exclusive license to Springer Nature Switzerland AG 2021
T. Myklebust, T. Stålhane, *Functional Safety and Proof of Compliance*,
https://doi.org/10.1007/978-3-030-86152-0_9

- Acceptance criteria: The criteria that a system or component must satisfy in order to be accepted by a user, customer, or other authorized entity, including performance requirements and essential conditions, which must be met before project deliverables are accepted
- Acceptance test: The test of a system or functional unit usually performed by the purchaser on his premises after installation with the participation of the vendor to ensure that the contractual requirements are met

The two tests most used are FAT (Factory Acceptance Test) and SAT (Site Acceptance Test). How the SAT is performed depends on the application domain. For an offshore installation, SAT is often just a repetition of the FAT at the customer's site and with the customer present. For railway systems, it is a two-step affair: first the developer runs the FAT to show that they have delivered as agreed in the contract and then the customer run his acceptance test—CAT. In many cases, the SAT is the first time all the units involved are tested together in a real environment even though the developers have run their tests in a simulated environment to test its behavior in an approximately real environment.

9.1.2 Factory Acceptance Test (FAT)

According to Ihle (2015), "all software shall be tested on the hardware as built for a particular installation.... This test ensures that all software and hardware in the positioning product works as specified." A simpler approach would be to say that the FAT shall test that the system complies with the specifications. This does not include only code. The following also need to be tested/checked:

- Installation description
- User manuals
- Manuals for operating personnel
- SRACs (safety-related application condition) and SecRACs (security RACs)

A test plan has to be established and, depending on the contract, agreed with the purchaser of the system. According to IEC TR 61511-4:2020—process industry—and IEC 62381:2012 (Factory Acceptance Test, Site Acceptance Test, and Site Integration Test), there should be a FAT plan. Important issues in this plan are:

- Resources allocated—time and personnel—both from the manufacturer and the purchasing company (if relevant).
- Tools and equipment needed. This includes available hardware such as computers, terminal, and network.
- When shall the FAT be executed.
- How will we handle errors discovered during FAT, e.g., resources allocated and deadline for corrections.

IEC 62381:2012, Annex A, has a template and a checklist for the FAT report. Some of the more important issues here are:

- System's alarm test
- Visualization and operation—graphic display arrangements
- Complex functionality and operation modes
- Integration of subsystems—needed to verify the interoperability of the systems involved

9.1.3 Site Acceptance Test (SAT)

While FAT is concerned with the system's specifications, SAT is concerned with the system's behavior under real operating conditions. Since a SAT involves a lot of information related to the company where it will be installed, a real SAT needs to be written by the customer or someone with in-depth knowledge of the customer's operations. This holds both for the test specifications, the test process, and the expected—acceptable—results.

The SAT should preferably be done by real users at the customer's site and with the customer's environment. This includes such things as the customer's network, drivers, PCs, and operating system. The number of terminals/PCs connected to the system and a realistic load situation are also important factors for the SAT. Some parts of the requirements may be difficult to reproduce during FAT. This holds, e.g., for the handling of dangerous situations during operation. For these cases, the FAT should use a "digital twin"—the generation or collection of digital data representing a physical object. Other simulation alternatives are also possible.

We need a different approach when developing software for a company that includes this software in its own products. A good example is the automotive industry which has made a standard for software development for vehicles—ISO 26262:2018. This standard does not use the terms SAT or acceptance testing. Instead, they have a section on "Correct implementation of the functional safety requirements at the vehicle level"—ISO 26262-4:2018, Table 13. This table introduces the concept "User Test Under Real-Life Conditions," described as "a long-term test and a user test under real-life conditions are similar to tests derived from field experience but use a larger sample size, normal users as testers, and are not bound to prior specified test scenarios, but performed under real-life conditions during everyday life. These tests can have limitations, if necessary, to ensure the safety of the testers, e.g., with additional safety measures or disabled actuators."

In similar ways, the EN 50128:2011, clause 8.4.5, has introduced the concept "Application Integration and Testing Acceptance." The standard states that "The Application Test Report shall document the correct and complete execution of tests defined in the Application Test Specification. The Application Preparation Verification Report shall check the completeness and correctness of tests performed on the complete installation."

9.1.4 Customer Acceptance Test (CAT)

According to Hein et al. (2015), Customer Acceptance Testing (CAT) is a means for ensuring the effectiveness of development efforts, i.e., are we developing something that the customers like? This is achieved by developing and testing prototypes to see how the product features fit customer explicit and implicit requirement. CAT is mainly used in market research where the quality of a prototype is tested subjectively by potential customers in order to make sure that product development efforts have moved the design and features toward customer requirements. We recommend that people with experience from marketing and surveys participate in this process to get information on what the *customers want*, not what the developers believe that they should wish to develop.

9.2 Overall Software Test Specification and the Use of Scripts

9.2.1 Test Specifications

The chapter on test specifications is mainly based on information found in ISO 26262-6 (automotive), EN 50128 (railway), IEC 61508-3 (generic functional safety), and IEEE 29119:2020 (generic AI-based testing).

We will use EN 50128 as our starting point since this is a well-known standard and it gives an extensive description of what a test specification shall contain. Note that, e.g., ISO 26262 only mentions test specifications once—in part 4—and otherwise just discusses test strategies. The same holds for IEC 615508. EN 50128 requires a software test specification to contain the following information:

(a) Test objectives
(b) Test cases, test data, and expected results
(c) Types of tests to be performed
(d) Test environment, tools, configuration, and programs
(e) Test criteria on which the completion of the test will be judged
(f) The criteria and degree of test coverage to be achieved
(g) The roles and responsibilities of the personnel involved in the test process
(h) The requirements which are covered by the test specification
(i) The selection and utilization of the software test equipment

The rest of this chapter will be organized according to this list of issues.

Test Objectives
The test objectives must be linked to one or more requirements. This includes both customer requirements and requirements imposed by the relevant standards. Tests not related to requirements are a waste of resources. The requirement may be either

functional—e.g., "input A will lead to response B" or "if error X occurs, the Y component will take control and lead the system to a safe state." Non-functional requirements are related to, e.g., safety, response time, user satisfaction, or use of computer resources—e.g., no command shall take more than X ms to handle.

Test Cases, Test Data, and Expected Results
A test case description consists of three components: test case, test data—input and system state data—and expected results. This part will be different for system with memory and system without it since systems without memory, e.g., a computational program, will react only on the current input, while a system with memory, e.g., an operating system, may react different to an input depending on its state—earlier inputs and operations. If the system operates in an interactive mode, we also need to specify the user or operator inputs.

Types of Tests to Be Performed
There are several ways to typify tests—e.g., "black box," "white box," functional testing, coverage testing, interface testing, and several others. An example of how to typify tests—from IEC 61508-3—is shown in the Table 9.1. The main thing in the overall software test specification is the test type since this decides what we will test and how we will test it.

Test Environment, Tools, Configuration, and Programs
The main purpose of this part of the test specification is to assure reproducibility and repeatability of the tests. In order to completely define the test environment, we need information on the system's version, the versions of the libraries used, and the version of the operating system. It will also be helpful to specify the computer, any hardware involved, and any network connections involved.

Jira provides good tools for test management (Atlassian 2020). The DZone home page (Dzone 2019) provides a list of ten popular software testing tools. The ones suitable for testing programming code are TestingWhiz, a test automation tool with the code-less scripting, HPE Unified Functional Testing (HP—UFT formerly QTP), and Tosca Test suite uses model-based test automation to automate software testing.

Test Criteria on Which the Completion of the Test Will Be Judged
A test campaign can end in two ways:

- All the tests run as expected and give the planned results.
- Some of the tests did not end as expected. Either the development company provides plans for how to correct the software to make the tests pass or they provide plans for a new analysis of the system and the system's requirements to check the correctness of the expected results.

The Criteria and Degree of Test Coverage to Be Achieved
For the IEC 61508-3, the relevant types of test coverage are described in Table B.2. Coverage is measured as a percentage of all possible items tested—e.g., if we want

Table 9.1   IEC 61508-3 next edition

Table A.5 – Software design and development – software module testing and integration

(See 7.4.7 and 7.4.8)

	Technique/Measure *	Ref.	SIL 1	SIL 2	SIL 3	SIL 4
1	Probabilistic testing	C.5.1	PR	R	R	R
2	Dynamic analysis and testing	B.6.5 Table B.2	R	HR	HR	HR
3	Data recording and analysis	C.5.2	HR	HR	HR	HR
4	Functional and black box testing	B.5.1 B.5.2 Table B.3	HR	HR	HR	HR
5	Performance testing	Table B.6	R	R	HR	HR
6	Model based testing	C.5.27	R	R	HR	HR
7	Interface testing	C.5.3	R	R	HR	HR
8	Regression testing	C.5.25	R	HR	HR	HR
9	Test management and automation tools	C.4.7	R	HR	HR	HR
10	Forward traceability between the software design specification and the module and integration test specifications	C.2.11	R	R	HR	HR
11	Formal verification	C.5.12	PR	PR	R	R

NOTE 1   Software module and integration testing are verification activities (see Table B.9).

NOTE 2   See Table C.5.

NOTE 3   Technique 9. Formal verification may reduce the amount and extent of module and integration testing required.

NOTE 4   The references (which are informative, not normative) "B.x.x.x", "C.x.x.x" in column 3 (Ref.) indicate detailed descriptions of techniques/measures given in Annexes B and C of IEC 61508-7.

*   Appropriate techniques/measures shall be selected according to the required systematic capability.

Test methods to be published 2022 or later. Text in yellow represents the foreseen changes from ed.2 to the next edition 3. © IEC

statement coverage, the measure will be 100*(number of tested statement/total number of statement). The standard considers structural coverage for

- Entry points
- Statements
- Branches
- Conditions

The standard provides more information in IEC 61508-7, C.5.8, which also shortly describes some other coverage measures (Table 9.2).

Note that the standard always requires 100% coverage for the selected measures. However, in most cases, an assessor will agree to a somewhat lower value—e.g., 99%—which has been accepted by one of the TUV organizations. If it is not possible to achieve 99–100% coverage, the reasons why this cannot be achieved should be documented in the test report.

Table 9.2   IEC 61508-3: next edition

Table B.2 – Dynamic analysis and testing

(Referenced by Tables A.5 and A.9)

	Technique/Measure *	Ref	SIL 1	SIL 2	SIL 3	SIL 4
1	Test case execution from boundary value analysis	C.5.4	R	HR	HR	HR
2	Test case execution from error guessing	C.5.5	R	R	R	R
3	Test case execution from error seeding	C.5.6	PR	R	R	R
4	Test case execution from model-based test case generation	C.5.27	R	R	HR	HR
5	Performance modelling	C.5.20	R	R	R	HR
6	Equivalence classes and input partition testing	C.5.7	R	R	R	HR
7a	Structural test coverage (entry points) 100 % **	C.5.8	HR	HR	HR	HR
7b	Structural test coverage (statements) 100 %**	C.5.8	R	HR	HR	HR
7c	Structural test coverage (branches) 100 %**	C.5.8	R	R	HR	HR
7d	Structural test coverage (conditions, MC/DC) 100 %**	C.5.8	R	R	R	HR

NOTE 1   The analysis for the test cases is at the subsystem level and is based on the specification and/or the specification and the code.

NOTE 2   See Table C.12.

NOTE 3   The references (which are informative, not normative) "B.x.x.x", "C.x.x.x" in column 3 (Ref.) indicate detailed descriptions of techniques/measures given in Annexes B and C of IEC 61508-7.

*   Appropriate techniques/measures shall be selected according to the required systematic capability.

**   Where 100 % coverage cannot be achieved (e.g. statement coverage of defensive code), an appropriate explanation should be given.

Test methods to be published 2022 or later. Text in yellow represents the foreseen changes from ed.2 to the next edition 3. © IEC

Table 9.3   Example test coverage report

Element	Missed Instructions	Cov.	Missed Branches	Cov.	Missed	Cxty	Missed	Lines	Missed	Methods	Missed	Classes
Palindrome.java		100%		100%	0	5	0	7	0	2	0	1
Total	0 of 38	100%	0 of 6	100%	0	5	0	7	0	2	0	1

There are several tools available that will show the test coverage of a test suite. Some of the popular test coverage tools are listed below:

- Java: Atlassian Clover, Cobertura, JaCoCo
- JavaScript: Istanbul, Blanket.js
- PHP: PHPUnit (PHPUnit is a programmer oriented testing framework for PHP)
- Python: Coverage.py
- Ruby: SimpleCov

The small table below shows an example of the results from applying JaCoCo to a small Java program (Table 9.3).

We see from the report above that the coverage is 100% for the component Palindrome.java.

At least Coverage.py has been criticized for not being of much help. To quote Batchelder (2007): "When your test coverage is less than 100%, coverage testing works well: it points you to the lines in your code that are never run, showing the way to new tests to write. The ultimate goal, of course, is to get your test coverage to 100%.

But then you have problems, because 100% test coverage doesn't really mean much. There are dozens of ways your code or your tests could still broken, but now you aren't getting any directions. The measurement coverage.py provides is more accurately called statement coverage, because it tells you which statements were executed. Statement coverage testing has taken you to the end of its road, and the bad news is, you aren't at your destination, but you've run out of road."

The tool is still useful when the assessor requires 100% test coverage. The criticism is about the usefulness of the measure when it comes to the tests' ability to discover errors. Thus, it should be used with some caution—quantum satis ("amount which is enough").

9.2.2   Scripts

According to Guru99 (2021), "Test Scripts are a line-by-line description containing the information about the system transactions that should be performed to validate the application or system under test. Test script should list out each step that should be taken with the expected results. This automation script helps software tester to test each level on a wide range of devices systematically. The test script must contain the actual entries to be executed, and the expected results."

Scripts are important in order to achieve test automation. There are several ways to categorize scripts. We will use the one presented by Hanna (2014):

- Linear Scripting Technique: Kent (2007) explains the idea behind linear technique, which is simply to set the test tool to the record mode while performing actions on the SUT (system under test).
- Structured and Shared Scripting Techniques: Structured scripting technique uses structured programming instructions, which can be either control structures or calling structures (Fewster 1999).
- Data-Driven Scripting Technique: Test data is stored in a separate data file instead of being tightly coupled to the test script itself. While performing tests, test data is read from the external data file.
- Keyword-Driven Scripting Technique: The business functions of the SUT are stored in a tabular format as well as in step-by-step instructions for each test case. Keyword-driven approach separates not only test data for the same test as in data-driven scripts but also special keywords for performing business function in the external file

In order to show you the flavor of scripts and to give you some ideas on what can be achieved using scripts, we will show you a few examples, written in Perl. For a short, good, and practical introduction to Perl, see Guru99 (2021). Perl is a high-level, general-purpose, interpreted, dynamic programming language. Perl stands for "Practical Extraction and Reporting Language" even though there is no acronym for Perl. It was introduced by Larry Wall in 1987. Perl was specially designed for text editing. But now, it is widely used for a variety of purposes including Linux system administration, network programming, web development, etc. Perl is a type-less, interpretative language and does not necessarily need a compiler.

In order to clarify the two examples, we will need a few Perl notations. Perl has many things in common with ordinary programming languages—e.g., Java, Pascal, and Ada. Perl has declarations, "for" and "while" loops and conditional statements—"if-else." In addition to the "for" statement, Perl also have a "foreach" statement. All scalars start with a $-sign. Thus, in the example below, $fromdate is a scalar. The @-sign indicates an array declaration. Thus, @parts and @opts are arrays. Perl has several useful built-in functions, e.g., the split function which splits the content of string into an array based on the delimiter provided to it. This function will also eliminate the delimiter from the string. In the example below, the first split is on "?". Thus, the array "parts" is set to the parts of "path" which are separated by "?". Next you split on "&" to get the options. To get the option values, you have to split each option on "=," and to get the list of option values, you have to split on ",".

Code 9.1 Example of Subroutine

```perl
#!/usr/bin/perl

use warnings;
use strict;

  sub getOptions{
      my ($path) = @_;
      my $fromdate = '2021-01-01';
      my @optvallist = ();
      my $option = 'default';

      my @parts=split(quotemeta('?'),$path);

      my @opts=split(quotemeta('&'),$parts[1]);
      foreach my $opt (@opts[0..$#opts]) {
        my @optparts = split ('=',$opt);
        if ($optparts[0] eq "fromdate"){
            $fromdate =$optparts[1];
        } else {
            $option = $optparts[0];
            my $optval = $optparts[1];
            @optvallist=split(',',$optval);
        }
      }
      return ($fromdate, $option, @optvallist);
  }
```

```
  my ($fromdate, $option, @optvallist) = getOptions("http://xxx.yy.
ch:8080/a/b/c.html?fromdate=2021-01-01&xxx=2,5,6&fromdate=2021-
07-01");
  print "Fromdate = $fromdate\n";
  print "Option = $option\n";
  print "Optvallist = @optvallist";
```

The last example is related to testing. It starts with initializing the t_path, the date, and the user's Perl commands—myperl. The "->" is used for dereferencing—i.e., to access a component in a package. The example below also uses hashes. Hash is defined by IEC 61784-3:2016 as a mathematical function that maps values from a (possibly very) large set of values into a (usually) smaller range of values. A hash can hold as many scalars as the array can hold. The only difference is we don't have any indices. Instead, we have keys and values. A hash is declared by using a "%" followed by the name of the hash.

Code 9.2 Example of Subroutine

```
@t_path = find_t_paths()
$date = get_date();$myperl = perl_command();

# Class interface
#
use Test::STDmaker

$std = new Test::STDmaker( @options ); # From File::Maker

$success = $std->check_db($std_pm);
@t_path = $std->find_t_paths(); $date = $std->get_date();
$myperl = $std->perl_command();

$std->tmake( @targets, \%options); $std->tmake(@targets);
$std->tmake( \%options);

# Internal (Private) Methods
#
$success = $std->build($std_driver_class); $success = $std->generate();
$success = $std->print($file_out);
```

9.3   Software/Hardware Integration Test Specification

The majority of the real-time safety-critical software systems interact with some hardware and the hardware used to execute the software. In most cases, the hardware involved is sensors giving information to the software system and actuators, allowing the software system to influence its environment. Thus, the way the software communicates with sensors and actuators—hardware—is of crucial impor- tance. Consequently, the test that checks that the integrator's software and hardware

Fig. 9.1 Example of an
ABS system

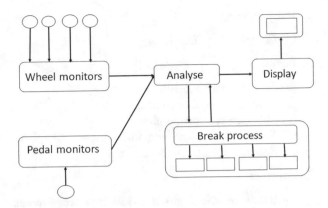

communicate correctly and are correctly integrated is important for safe operations. The Fig. 9.1 shows an example of system with hardware-software integration—an ABS (Anti-lock Braking System/Automatic Braking System) system.

The starting point for hardware-software integration is the hardware-software architecture that describes which part is needed to fulfill which requirement. The architecture needs to show how hardware, company-developed software, and pre-developed software are integrated. In addition, the architecture should use a well-proven pattern—e.g., the observe-react pattern.

The software architecture shall also consider the feasibility of achieving the software requirements specification at the required software safety integrity level. As always, it is important that the software architecture minimizes the size and complexity of the safety part of the application.

This section is based on input from three important standards—IEC 61508, ISO 26262, and EN 50128. EN 50128 has the most detailed requirements related to HW/SW integration. Except for this first one—IEC 61508, which is a generic standard—the two other standards are specifically concerned with software/hardware communication: ISO 26262 for cars and EN 50128 for trains.

On HW/SW Integration Testing in General
As with most other processes, the integration test can be described by the following five pieces of information and activities:

- Entry criteria
- What must be done before we can start, inputs
- What do we need to get started, what are we going to do, and when are we finished
- The exit criteria and what are the results
- The output

The following list is copied from Guru99. We have added some issues in *italics* to cater to the needs of safety-critical systems:

1. Entry criteria:

   (a) Completion of unit testing

2. Inputs:

   (a) Software requirements
   (b) Software design document
   (c) Software verification plan
   (d) *Software/hardware integration test specification*
   (e) Software integration documents

3. Activities:

   (a) Based on the high- and low-level requirements, create test cases and procedure.
   (b) Combine low-level modules' builds that implement a common functionality.
   (c) Develop a test harness.
   (d) Test the build.
   (e) Once the test is passed, the build is combined with other builds and tested until the system is integrated as a whole.
   (f) Re-execute all the tests on the target processor-based platform, and obtain the results.

4. Exit criteria:

   (a) Successful completion of the integration of the software module on the target hardware
   (b) Correct performance of the software according to the requirements specified, *including safety requirements*

5. Outputs

   (a) Integration test reports
   (b) Software test cases and procedures
   (c) *Safety integration report and/or software/hardware integration test report*

It is up to each company to define what is sufficient integration documents, what is a good test harness, how the build shall be tested, and so on. However, the developing company needs to decide what should be sufficient integration documentation and why it is so—either by presenting a set of arguments or by referring to applicable standards. And, as always, the assessor has the final word.

Some of the activities listed above are either obvious or already done before we arrive at the stage where hardware and software should be integrated. This goes for 1; 2a, 2b, and 2c; 3d; 4a and 4b; and 5b. These activities will thus not be discussed any further.

On Safety Requirements

The challenge when it comes to safety requirements is that they are non-functional. Thus, they are more related to the development process—what we have done—than to how we did it. There is no way to ensure that the system is absolutely safe—neither before or after installation. The best we can do is to check that all process requirements are met as specified in the relevant standards. We can use IEC 61508-3:2010, Annex A, as an example. Some of the requirements here are:

- The use of formal methods
- Forward and backward traceability
- The use of structured methods
- Defensive programming

The list goes on through Annexes A.1 to A.10. The ISO 26262 likewise has 15 tables with requirements—Table 1 to Table 15. These requirements are mostly similar to the ones used in IEC 61508-3. Most of the safety requirements are not fit for testing. Thus, we need code inspection and inspection of all process documentation if we want to check for safety before the HW/SW integration starts.

IEC 61508-3

IEC 61508-3 considers the following issues to be important for the integration test:

- Completeness and correctness according to the design specifications
- Repeatability, e.g., in the case of software or hardware modifications
- Precise definition of the configuration

In addition, IEC 61508-3 highly recommends functional black box testing for all SIL values and performance testing for systems with SIL 3 or SIL 4 requirements.

The manufacturer must develop the integration test during software design and development. We will assume that the part of the hardware that will communicate with the software—the hardware-software interface—is already defined. If the interface changes, the persons responsible for the integration tests need to be informed, and the tests must be updated if necessary. We need to combine the software and hardware in the safety-related programmable electronics to ensure hardware-software compatibility and meet the intended safety integrity-level requirements. If we include any pre-developed software, it is important to verify that this software is developed at least to the same SIL as the one we have developed. We also need to check that the pre-developed software can handle relevant hardware and software errors.

As for all test specifications, we need to specify the hardware to be used, the input to be used in the tests, and the expected results. The integration test specification need to distinguish between the following activities:

- Merging of the software system onto the target hardware
- Integration, i.e., adding input and output devices such as sensors and actuators
- Applying the resulting safety-related control system to the equipment it will control

If we compare the activities identified in the Guru99 list with the requirements specified by IEC 61508-3, we see that the standard does not touch issues 1, 2, and

5. What the standard misses are the SW integration document and the integration test report—both of these are documents that need to be part of any integration test.

ISO 26262-2:2018

A more extensive description of how to develop a hardware-software integration test can be found in ISO 26262-2. This standard requires that the following analysis shall be performed. Analysis of:

- Requirements
- Boundary values
- External and internal interfaces
- Environmental conditions and operational use cases
- Functional dependencies

ISO 26262-2 has some specific test requirements in that you need to do fault injection tests, back-to-back tests, and performance tests. The only places where this standard explicitly agrees with the Guru99 list are issues 2a and 3a.

In addition, ISO 26262-2 requires that the integration test shall check for timing of the safety mechanisms, hardware fault detection, and robustness. Which hardware errors will the system be able to diagnose, and which error will it be able to handle? If we look back at Fig. 9.1, important questions related to safety criticality would be, e.g.:

- What happens if you lose contact with the pedal monitor or one or more of the wheel monitors?
- How will a display error influence the driver?
- What will happen if the break process hangs?

In order to get answers to these and similar questions, we need to have the real hardware available so that we can do testing using fault injections.

EN 50128:2011/AC2020

The EN 50128 has the most extensive descriptions of the hardware-software integration tests but on a high level of abstraction. Of special value is their list of what an integration test specification shall show—see Sect. 6.1.4.4. The following is a slightly edited version of this list.

- Test objectives—what do we want to achieve with the test, e.g., related to a requirement? The planned test coverage for the software should also be included here.
- Test cases, test data, and expected results.
- Types of tests to be performed—e.g., functional test, error recovery test, or timing test.
- Test environment, tools, configuration, and programs. For example, is it OK to run the test on a digital twin or do we need to run it on the real system?
- Test criteria and acceptance criteria on which the completion of the test will be judged. This is strongly related to test objectives, test type, and test environment.

Section 7.3—Architecture and Design—contains important information in Sect. 7.3.3. The documents needed for integration test, e.g., its specification, should be written here. The EN 50128 also requires that the development organization should have an integrator role which is responsible for developing the integration test and writing the integration test report. Related to the Guru99 list, the EN 50128 contains issues 2c and 2b plus all 4 and 5 issues.

Summing It All Up
For all the standards involved here, following Guru99's list will do. However, for some of the standards, it might be an over-kill. On the other hand, several of the activities mention in the list are done for most software development projects but are not considered as part of the integration test. This goes for, e.g., the activities in issues 1 and 2. ISO 26262-2 is the standard with the most detailed integration test requirements since it contains requirements for fault injection tests, back-to-back tests, and performance tests. These tests should be used regardless of the relevant standards.

9.4 Software Quality Assurance Verification Report

The purpose of the software quality assurance verification report is to show that the quality assurance processes for the development project has been done as required by the official standards, the company's internal standards, and the customer. The quality assurance process should be stated in the software quality assurance plan (SQAP) plus the project's safety plan. In addition, we should consider standards such as the IATF 16949:2016 standard for the automotive domain and, e.g., ISO/TS 22163:2917 for the railway domain. This is a supplemental standard and is used in conjunction with several ISO standards, e.g., ISO 9000:2015, ISO 9001:2015, and ISO 9004:2018.

The purpose of the software quality assurance verification report is to show that the developer (supplier) has done the following, as stated by ISO 9000-3: 2018: "The supplier shall establish, document and maintain a quality system as a means of ensuring that product conforms to specified requirements. The supplier shall prepare a quality manual covering the requirements of this International Standard. The quality manual shall include or make reference to the quality system procedures and outline the structure of the documentation used in the quality system."

IEC 61508-3 contains three important annexes (A, B, and C) that describe the techniques and measures that should be applied during software development. Only Annex A is mandatory, but several assessors also require adherence to Annex B. Annex C is informative—i.e., not required but still quite useful. In addition, IEC 61508-3:2010, clause 6.2.17, states that the development organization should have "an appropriate quality management system."

First and foremost, we need:

- The software quality assurance plan (SQAP)
- A description of the verification methodology, if this is not part of another document

In addition, depending on the domain, we need the following documents:

- ISO 9001, Quality management systems—Requirements
- ISO 9000-3, Guidelines for the application of ISO 9001 to development, supply, installation, and maintenance of computer software
- IEEE 730, IEEE Standard for Software Quality Assurance Plans
- ISO 25001, Systems and software engineering—Systems and software Quality Requirements and Evaluation (SQuaRE)—Planning and management
- IEC 61508-3, Annexes A, B, and C—a generic software standard
- EN 50128—a railroad standard
- ISO 26262-6—an automotive standard
- The project's safety plan
- The project's quality plan

At the top level, quality assurance is simple. The supplier tells you what they will do to ensure that you get what you have ordered—the quality assurance management system—and when they deliver the product, they produce a verification report, showing that they have done what they promised. The verification is done by checking the "paper trail" of each planned process activity. For example, if the contract states that all code shall be reviewed, it must be possible to access review reports for all code—what was reviewed, how was it done, who did it, and what was the result.

Note that ISO 9001 auditors are willing to accept whiteboard snapshots of the discussions if it includes the date and a list of participants as proof of conformance for an activity. At least some IEC 61508 safety assessors have confirmed that they will do the same.

According to ISO 9001:2015, the quality management system documentation shall include:

- Documented statements of a quality policy and quality objectives
- A quality manual
- Documented procedures required
- Documents needed by the organization to ensure the effective planning, operation, and control of its processes
- Records required by the standard (see ISO 9001:2015, 4.2.4)

In addition, the quality plan should be in accordance with ISO/IEC 90003:2014 "Software engineering—Guidelines for the application of ISO 9001:2008 to computer software." Note that ISO 9001 requires that the quality system should be reviewed by management at planned intervals.

The verification process must assure that:

- All process activities are done according to the quality plan
- All process activities leave a "paper trail"

- The activity descriptions show that the activities are done as described in the quality plan

In all cases, it is important to consider the consistency and the complementarity of the chosen methods, languages, and tools plus the developers' understanding of and competence in the chosen methods, languages, and tools. This might be a challenge, for instance, for formal methods. What is required according to the annexes for IEC 61508-3:2010 will depend on the SIL value. Moving from SIL 2 to SIL 3 will not change many requirements—only 12 out of 91. The big change occurs when going from SIL 3 to SIL 4—40 out of 91 requirements. The requirements that change from R to HR for SIL 2-SIL 3 in Tables A and B are as follows:

- A2: fault detection and semi-formal methods
- A4: defensive programming
- A5. testing
- A7: modelling
- A10: functional safety assessment
- B2: structured testing and boundary value analysis
- B8: inspection
- B9: complexity control

For an agile development process, we may, in addition, need to look at the paper trail of the alongside engineering activities.

For the safety case, the main purpose of the software quality assurance verification report is to support the other evidences. For example, if we claim that we have achieved full traceability, the software quality assurance verification report will support this claim by showing that the process to ensure traceability has been done the right way.

If they are not referred to in another document, the specialized part needs the following subsections:

1. Coding standard(s) used
2. Code baseline

Railroad Example

IRIS (International Railway Industry Standard) certification bodies certify manufacturers of railway components. The aim of IRIS is to develop and implement a global system for the evaluation of companies supplying components to the railway industry with uniform language, assessment guidelines, and mutual acceptance of audits, which will create a high level of transparency throughout the supply chain. See also RINA (2020).

IRIS, together with ISO 9001:2015, defines the business management system requirements for the design and development, manufacturing, maintenance, and, when applicable, installation, customer service of rolling stock, and signalling-related products. If the Quality Management Report (QMR) is a separate document, we should include an overview of the project and product, purpose, intended audience, scope, structure, and evolution of the QMR. This is especially important if an agile approach is used since more updates of the QMR and the safety case are foreseen. If not included in a separate document, definitions, acronyms, and references should be included (Fig. 9.2).

All the quality factors are important. We will here just have a quick look at maintainability. Definitions of the other quality characteristics are also available.

According to the SQuaRE model, maintainability consists of subjective or objective (metrics) measures of the degree to which:

- A system or computer program is composed of discrete components such that a change to one component has minimal impact on other components—modularity
- An asset can be used in more than one system or in building other assets—reusability
- It is possible to assess the impact on a product or system of an intended change to one or more of its parts, or to diagnose a product for deficiencies or causes of failures, or to identify parts to be modified—analyzability
- A product or system can be effectively and efficiently modified without introducing defects or degrading existing product quality—modifiability
- Test criteria can be established for a system, product, or component and tests can be performed to determine whether those criteria have been met—testability

9.5   Software Architecture and Design Verification Report

Two questions are important: does the architecture (1) meet the customer's requirements and (2) comply with the relevant standards.

The first issue is handled by the previous section—software architecture specification. The second issue is handled in this section. In order to handle this second issue, we need to look at two pieces of information—the architecture process and how the architecture should be evaluated. Software architecture verification should check that the software architecture design fulfills the software requirements specification and ensure that the integration tests specified in the software architecture are adequate. We need to check the software architecture design, see how it handles the architectural attributes of the subsystems, and check for incompatibilities between design and specification.

This section is based on the two standards ISO/IEC/IEEE 42030:2019 "Software, systems and enterprise—Evaluation framework" and ISO/IEC/IEEE 42020:2019 "Software, systems and enterprise—Architecture processes." In addition, we have used IEC 61508-3, Sect. 7.4.3.2, on software architecture design.

According to the software architecture specification section, the following issues are important for SIL 3 architectures:

- Fault detection and graceful degradation
- Modular approach

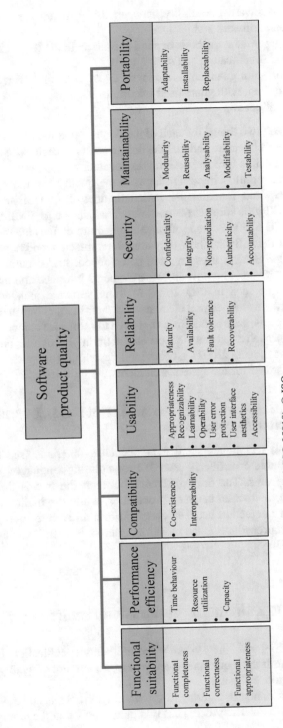

Fig. 9.2 The SQuaRE software product quality model (ISO 25010). © ISO

- Forward and backward traceability between the software safety requirements specification and software architecture
- Structured diagrammatic methods or semi-formal methods (e.g., UML)
- Computer-aided specification and design tools
- Cyclic behavior, with guaranteed maximum cycle time, time-triggered architecture, or event-driven, with guaranteed maximum response time
- Static resource allocation

We need to check two things for each of the issues—that the components needed are included in the architecture and that the design caters to each issue in an efficient way. This should include how the design was decided.

The architecture report is important input to the quality assurance, the requirements—are all required functionality included, does it fit with the application architecture and will it be suitable for the safety analysis—e.g., FMEA.

According to IEC 61508-3, "the software architecture defines the major elements and subsystems of the software, how they are interconnected, and how the required attributes, particularly safety integrity, will be achieved. It also defines the overall behaviour of the software, and how software elements interface and interact."

We find the same idea in ISO 26262-6: "The software architectural design represents the software architectural elements and their interactions in a hierarchical structure. Static aspects, such as interfaces between the software components, as well as dynamic aspects, such as process sequences and timing behaviour, are described."

According to Bachmann (2002), a behavioral description supports exploring the range of possible orders for the system's elements:

- Interactions
- Opportunities for concurrency (see also Annex F of IEC 61508-3:2010)
- Time-based interaction dependencies

Two (traditional) techniques stand out—sequence diagrams (see IEC 61131-3:2013 "Programmable controllers—Part 3: Programming languages") and procedure sequence diagrams. The only difference between the two is that procedure diagrams show the interaction between procedures, while the traditional sequence diagrams show the interaction between classes. From an architectural perspective, interaction between procedures is the most important one when it comes to describing architectural behavior.

9.6   Software Requirements Verification Report

For software development, the system's requirements specification is the most important document. If the requirements are wrong, no amount of quality assurance, testing, and inspection will help.

There are several standards catering to software requirements and how to collect, validate, and verify them. The process has two parts: show that the requirements are

(1) as the customer wants them and are (2) implemented as specified. The latter part again has two components—the quality of the process and the realized functional and non-functional software requirements. These issues are handled in Sect. 9.7.

- ISO/IEEE 29148:2018: Systems and software engineering—Life cycle processes—Requirements engineering
- IEEE Std. 1012:2016: Standard for System, Software, and Hardware Verification and Validation
- IEC 61508-3:2016: Functional safety of electrical/electronic/programmable electronic safety-related systems—Part 3: Software Engineering

The safety requirements are outlined and structured in IEC 61508-3, Annexes A and B.

First and foremost, we need a verification plan. IEEE 1012 suggests the following structure:

1. Purpose
2. Referenced documents
3. Definitions
4. V&V overview
5. V&V processes
6. V&V reporting requirements
7. V&V administrative requirements
8. V&V test documentation requirements

The main needs for a requirement are that it must:

- Be possible to verify the requirement. It must be possible to collect evidence that proves that the system satisfies the specified requirement. Verifiability is enhanced when the requirement is measurable.
- Be useful for the system when solving a stakeholder problem or to achieve a stakeholder objective.
- Be qualified by measurable conditions and bounded by constraints. This is important for verification and validation.
- Define system performance when used by a specific stakeholder or the corresponding capability of the system.
- Not define the capability of the user, operator, or other stakeholder.

According to IEEE 1012, the requirement evaluation has five steps—checking for (1) correctness, (2) consistency, (3) completeness, (4) readability, and (5) testability. The requirements verification report must, for each requirement, show:

- The requirement
- Which part of the system is involved in realizing the requirement
- How it was verified—e.g., the method(s) used to implement the five evaluation steps in IEEE 1012

9.7 Overall Software Test Report

For generic products, the overall software test report is the final test before the product is shipped from the manufacturer—also known as FAT (Factory Acceptance Test). The goal of this test is to show that all explicit and implicit requirements have been satisfied. EN 50128 states that "The objective of FAT is to test the devices of a safety instrumented system to ensure that the requirements defined in the software requirement specification are met. The FAT is sometimes referred to as an integration test and can be part of the validation. Testing of field elements together with a logic solver can be recommended when there needs to be a high confidence in operation prior to final installation, e.g., subsea applications."

IEC 61511-1:2016 has a good set of requirements for FAT. It starts by requiring that the planning for a FAT shall specify the types of tests to be performed including black box system functionality tests; performance tests; internal checks; performance tests; environmental tests; interface testing; testing in degraded or faulted condition; exception testing; testing for safe reaction in the case of power failure (including restart after power restored); and application of the SIS maintenance and operating manuals. In addition, IEC 62381:2012 should also be consulted.

For specific applications—e.g., railroad trackside equipment—the Site Acceptance Test (SAT) will be the final test. For several domains, the product's safety case and the independent assessor's report will also be needed before shipping the product.

According to EN 50128, the tester is responsible for writing the test specifications, running the tests, and writing the test report. However, IEC 61508-1, clause 6.2.3, only requires that "All persons, departments and organizations responsible for carrying out activities in the applicable overall, E/E/PE system or software safety lifecycle phases shall be identified, and their responsibilities shall be fully and clearly communicated to them."

Last, but not least, beware the difference between validation and verification. According to IEEE:

- Validation: The assurance that a product, service, or system meets the needs of the stakeholders.
- Verification: The evaluation of whether or not a product, service, or system complies with a regulation, requirement, specification, or imposed condition. It is often an internal process.

In this section, we are discussing validation.

As standards and guidelines, we have used the following documents:

- IEEE 29119-3:2013: IEEE Standard for Software Testing—Test Documentation
- IEEE Std. 829:2008: IEEE Standard for Software Test Documentation
- IEEE 829:2008: Test Summary Report Template

Use cases, hazard and safety stories, etc. (see Stålhane and Myklebust 2018) are input to the developers' requirements and are not needed here. We will assume that the requirements specification is correct. The following documents are important input to the overall software test report:

- The requirements specification, including the operational environment.
- The developer's test plan—who will test what and when? Here we also need to know how the testes were developed—e.g., how was each requirement mapped onto one or more tests?
- The test cases together with expected results. Are all requirements being tested?
- The test log summary—the results of the tests. Did the tests run as expected?

The "Overall software test report" part of the safety case must confirm that testing is done according to the test plan. According to IEEE 829, a test case should have the following contents:

1. Test case identifier.
2. Test item—the software to be tested by this test case.
3. Input specifications—the inputs needed, e.g., input data.
4. Output specifications—the expected results.
5. Environmental needs—where is the software run and which services must be available on the computer?
6. Special procedural requirements.
7. Dependencies between test cases—are there other test cases that must have run correctly before this test can run?

The output from the test process is the test report. This report needs to describe:

- Which requirements have been tested in full, partly, or not at all? What are the consequences for the users if the system is shipped in its present state?
- Which tests did not give the expected/specified result? Note that in this case there are two alternatives—(1) the system has a fault or (2) the specified result is wrong. For case (1), we need to consider the consequences of shipping the system "as is" and fix the problems in the next release. In order to decide on this, we need to analyze the consequences for the system's users.

9.8 Software Validation Report

9.8.1 Verification vs. Validation

The first thing we need to clarify is the difference between validation and verification. According to Easterbrook (2010):

- Validation: Are we building the right system?
- Verification: Are we building the system right?

Thus, validation is concerned with checking that the system will meet the customer's actual needs, while verification is concerned with whether the system

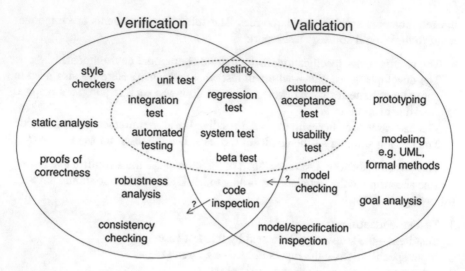

Fig. 9.3 Verification and validation—an overview of techniques and methods

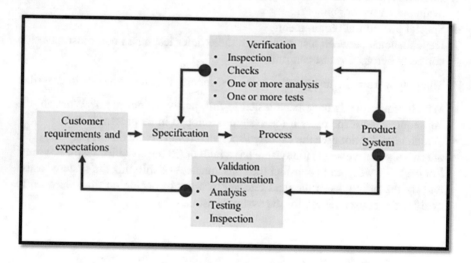

Fig. 9.4 Verification and validation related to end customer

is well-engineered, error-free, and so on. Verification will help to determine whether the software is of high quality, but it will not ensure that the system is useful. The following diagram, also taken from Easterbrook (2010), is useful. Note that everything inside the red circle belongs to the testing activities (Fig. 9.3).

If we instead apply a view related to specification and validation related to end customer, the following diagram is useful (Fig. 9.4).

If we stick to the model referred to in Easterbrook (2010), there are some important things to consider:

- There are several test activities that belong to both verification and validation—e.g., regression test and system test.
- While, e.g., unit testing and integration testing belong to verification, usability tests and customer acceptance tests belong to validation.

If we, instead, use the IEEE standard glossary—IEEE 610, we find the following definitions.

- Verification. The process of evaluating a system or component to determine whether the products of a given development phase satisfy the conditions imposed at the start of that phase.
- Validation. The process of evaluating a system or component during or at the end of the development process to determine whether it satisfies specified requirements. Note that "specified requirements" does not necessarily mean "all needed specified requirements."

If we look at the definitions used in the three standards IEC 61508, ISO 26262, and EN 50128, we see that IEC 61508 and EN 50128 are quite similar: EN 50128 and IEC 61508-3: Safety validation is to ensure that the integrated system complies with the software safety requirements specification at the required safety integrity level. ISO 26262 differs from the two previously mentioned standards since it defines validation as follows: Validation aims to provide evidence of appropriateness for the intended use and aims to confirm the adequacy of the safety measures for a class of vehicles. The difference is important, both for developers, verifiers, and validators. For EN 50128 and IEC 61508, the agreed-upon set of requirements is enough for validation—what you specify is what you get. If you find out later that you should have written different requirements, that's OK—write new requirements and pay a new bill. The probability of this rather unfortunate situation will be considerably reduced if you use an agile approach.

For ISO 26262, the situation is different in that it requires the validator to confirm the adequacy of the safety measures for a class of vehicles. As a consequence, there is no complete list of requirements, just a need to validate the fitness for use. According to Tedder (2018), fitness for use can be achieved by considering the following four advices:

- Capture the voice of the customer.
- Go to the where the real work happens—before and after.
- Take an Agile approach to solution delivery.
- Make sure the solution includes training and communication.

The result of applying these four advices must be documented and will serve as a starting point, both for validation and for acceptance criteria.

In our opinion, the IEEE 610 definition is more in line with the way the terms are used in software development companies. Thus, we will use this definition in the rest of this section. Just remember, there are other definitions also used "out there."

Validation of safety-related software shall include choice of techniques—manual or automated, static or dynamic, and analytical or statistical—and acceptance criteria, based on objective factors. See also Sect. 4.5—Safety Techniques and Measures. We will discuss each item related to a validation report in the subsections below.

If we use an agile development process, we should include alongside engineering in the process. This implies that we will have a validation process after each sprint to take care of the system's safety—the RAMS process. See also Sect. 2.6—Alongside Engineering—in this book. Note that the RAMS process can be applied in two ways—(1) at the end of a sprint to validate what the sprint has produced and (2) during a sprint to validate what the previous sprint produced.

9.8.2   Special Problems Related to Self-Driving Vehicles

Development of software for self-driving cars follows ISO 26262. D. Tedder's advice (Tedder 2018) about going to where the real work happens—i.e., out in the streets—is difficult. An alternative is to describe a set of scenarios that should be used in tests. However, it is not possible to test every possible scenario. Some of the challenges are that it is difficult to:

- Statistically demonstrate avoidance of unreasonable risk and a positive risk balance
- Consider previously unseen scenarios involving a single automated driving system and scenarios related to interactions between automated driving systems
- Validate system and elements that are likely to face software updates over its lifetime
- Validate systems that are based on automated driving systems, relying on machine learning algorithms

Even so, there are several ways that we can validate car control software—e.g., statistical testing, scenario-based testing, and field monitoring of the system in use.

- Statistical gray box testing: the scenarios used are taken from the ODD. Thus, it is possible to test the system using real-world scenarios. However, since we have to do some random selections of scenarios, we cannot be sure that all critical scenarios are covered.
- Scenario-based testing: testing using real-world scenarios is an important way to test a car control system. However, not all types of scenarios can be easily constructed—e.g., complex, heavy traffic or near misses in heavy rain.

Field monitoring: observing the behavior and collecting data from real-life experiences is surely the best way to test a car control system. This will help us by providing real-life situations which we later can analyze and use to develop software system improvements. Whether this is doable or not will depend on the level of safety that we have achieved before letting the cars loose into the street traffic.

9.8.3 Relation to the Software Validation Plan

The software validation plan needs to specify:

- How each requirement in the SRS will be validated—input, system state, and results.
- How each activity shall leave the required paper trail. This holds for data used as input to tests, the results obtained, persons participating, and resources used.
- If required by the standard, we need to identify the validator and how to ensure his independence.

If we are using the EN 50128, we need to keep in mind that the validator may require additional tests. As stated by clause 7.7.4.3 of the standard: "the Validator shall specify and perform supplementary tests on his discretion or have them performed by the Tester. While the Overall Software Tests are mainly based on the structure of the Software Requirements Specification, the added value the Validator shall contribute, are tests which stress the system by complex scenarios reflecting the actual needs of the user."

If the system is supposed to be resilient to one or more error situations, this should be noted in the safety plan.

9.8.4 Techniques and Methods

There are several techniques and methods available for software validation—see, e.g., IEC 61508-3, Table A.7, Software aspects of system safety validation; ISO 26262-6, Table 8; and EN 50128, Table A.5, Verification and testing. All the three standards also contain requirements for test coverage, e.g., for code or entry-exit points. The validation report must specify:

- The computer environment used for validation. This includes, but is not limited to, computer type; available memory; relevant systems installed, such as compilers and operating system; and, if needed, available networks.
- The method and tools used to achieve and compute the necessary coverage criteria.
- The coverage achieved.

Resilient systems give a special challenge for testing. Resilience is "the ability of a system to handle unexpected situations and recover." Thus, resilience engineering focuses on how to detect early and avoid (if possible); handle, ideally without disruption but may be reduced state; and recover—i.e., fail fast and get back on track. In order to validate a system's resilience, we need to do the following:

- Use a fault injection to cause the system to fail in a way that it is supposed to handle—a resilient situation.

- Observe the system's behavior—is it consistent with the behavior required by the specification?

9.8.5   Acceptance Criteria

The pass/fail criteria for accomplishing software validation shall include the required input and output—sequences and values—and any other acceptance criteria, such as memory usage, timing, and achieved coverage as required. In addition, we should report the resources used—personnel, tools and equipment, and hours used. This information is not part of proof of compliance but is still important since it helps to build trust to the results.

9.8.6   Error Handling

The company needs to have a plan for handling errors that may occur during validation. This plan must contain:

- Who shall report the errors that occur during validation
- Who is responsible for errors correction and revalidation—e.g., regression testing

If there is only one validator, he is responsible for reporting to the person who is responsible for system updates. He will either update the system or—in some cases—update the expected result. If there are several persons involved in the validation, there must be an agreement on who shall report validation errors to the persons responsible for updating the system or the tests.

9.8.7   Validation of AI-/ML-Based Systems

We have not found any complete description on how to test AI or ML systems. This is an indication that this field is far from ready for validation of industrial, safety-critical systems. What follows below is a short summary of two important papers in the area (Hand and Khan (2020) and Tao et al. (2019)). We should probably abide by the advice from Xie et al. (2011): "formal proofs of an algorithm's optimal quality do not guarantee that an application implements or uses the algorithm correctly, and thus software testing is necessary."

According to Hand and Khan (2020), the basic question related to using an AI system is "Can you trust your AI? How do you know it is doing what you want it to do?" This rather high-level question can be split up into several sub-questions, e.g.:

1. Has the objective been properly formulated?
2. Is the AI system free of software bugs?
3. Is the AI system based on properly representative data?

4. Can the AI system cope with anomalies and inevitable data glitches?
5. Is the AI system sufficiently accurate?

Question 1 involves mapping real-world questions to a description that can be described in a programming language. This question is context dependent, and there is little we can say in general about it. Questions 2–5 depend on data and algorithms. The old saying "garbage in, garbage out" also holds for AI and ML. An additional problem is the environment changes, and as the environment changes, we may need to change input data to a machine learning system. Thus, the validation of an AI or ML system needs to consider the environment as it is now and the environment in which the system will operate in say 4 years. An additional problem is that the data we will use might come from several sources and thus vary in reliability and quality. As a last worry, the European Union's General Data Protection Regulation (clause 71) says "automatic processing of data should be subject to suitable safeguards, which should include the right to obtain an explanation of the decision reached." This will be doable for most AI systems but way out of reach for most ML systems.

An alternative approach has been suggested by Tao (2019), which claims that AI software testing can be performed by using several perspectives:

- Classification-based AI software testing. We need to assure adequate testing coverage of input data, environment descriptions, and corresponding outputs.
- Metamorphic testing—extensively described in ISO/IEC TR 29119-11: 2020, clause 8.4.
- Model-based AI software testing. One or more learning models and data models are updated to serve as AI test models allowing testing and operations for quality assessment of test data.
- Learning-based AI software testing. Selected machine learning models are used to learn from crowdsourced testers.
- Rule-based AI software testing—pre-defined expert-based rules are used in AI test generation and validation.

In addition to the five approaches mentioned in the bullet list above, ISO/IEC TR 29119-11, clause 8, also recommend the following approaches for testing of AI and ML systems:

- Combinatorial testing: A combination of interest is constructed by first identifying inputs and environment conditions and the values these parameters can take. Combinations are constructed by sampling the input space and the environmental space. The input space may be reduced by using the General Morphological Analysis (GMA) approach (Ritchey 2013).
- Back to back testing: Another, already existing version of the system, using a different programming language or a different approach, is used as an oracle that is later used to check the test results for the system under test.
- A/B testing is about comparing two systems to see which one is the best. In some sense, it is just back-to-back testing, except that there is no "gold" version.

- Exploratory testing: Testing that allows for creativity and rapid execution of tests. Basically, it is just "Give it some input and see what happens." Thus, way to test is beneficial when specifications are poor or lean, e.g., in agile development.

Note that there is nothing new here—all of these approaches are already used for regular software systems. What is new is that the standard recommends them for AI and ML systems.

Scenarios play an important role in testing of, e.g., autonomous vehicles. These scenarios can be built based on information from several sources. ISO/IEC TR 29119-11, clause 10.2, identifies the following approaches: scenarios based on:

- What the system is supposed to do
- User reports
- Issues reported automatically, e.g., for autonomous vehicles
- Accident reports
- Reports from insurance companies—e.g., for autonomous cars
- Regulatory data collected through legislation
- Testing at various levels (e.g., test failures or anomalies on the test track or on real roads could generate interesting test scenarios for an autonomous car at other test levels, while a sample of, e.g., test scenarios run on the virtual test environment should also be run on real roads to validate representativeness of the virtual test environment)

9.8.8   Data Considerations

The choice and use of data are an important part of validation. This holds both for AI, ML, and traditional software.

For traditional software, the process is simple. When we chose which data to use for each test, we need to specify (1) which requirement we are testing and (2) why we chose this particular dataset to test it. For example, IEC 61508-3, Tables A5 and A7, require black box testing for all SIL values. It is important to know which data were used for black box testing and why did we choose these data instead of other ones.

Data plays an important role in ML since they are used to encode the requirements which will be used to instantiate the ML system. The data requirements shall ensure that it is possible to meet the safety requirements. The most important data requirements are the safety requirements related to the description of the system environment. Hawkins et al. (2021) have defined a process that can be used to identify the data needed to instantiate the ML system—see Fig. 9.5.

The Integrated ML Data Argument Pattern is an argument that "The data used to develop and verify MLM (Multilevel Marketing) is sufficient." The first two levels of the argument structure are shown below (Hawkins et al. 2021) (Fig. 9.6).

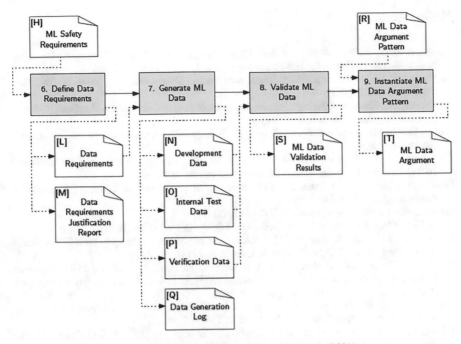

Fig. 9.5 ML data requirements assurance process. © Hawkins et al. (2021)

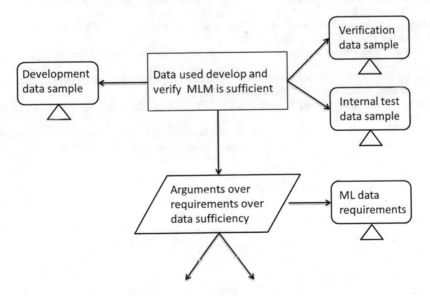

Fig. 9.6 The first two levels of the data sufficiency argument

References

Atlassian: Using JIRA Software for Test Case Management. https://confluence.atlassian.com/jirakb/using-jira-software-for-test-case-management-136872198.html (2020)

Bachmann, F., et al.: Documenting software architecture: Documenting behavior. In: Technical Note CMU/SEI-2002-TN-001 (Jan 2002)

Batchelder, N.: Flaws in Coverage Measurement. https://nedbatchelder.com/blog/200710/flaws_in_coverage_measurement.html (2007)

Dzone: Top 10 Most Popular Software Testing Tools. https://dzone.com/articles/top-10-automated-software-testing-tools (2019)

Easterbrook, S: The Difference Between Verification and Validation. www.easterbrook.ca/steve/2010/11/the-difference-between-verification-and-validation (2010)

Fewster, M.: Software Test Automation: Effective Use of Test Execution Tools. Addison-Wesley Professional, Reading, MA (1999)

Guru99.: www.guru99.com/perl-tutorials.html (2021)

Hand, D.J., Khan, S.: Validating and Verifying AI Systems. Cell Press (2020)

Hanna, M., El-Haggar, N., Mostafa, M.-S.: A review of scripting techniques used in automated software testing. Int. J. Adv. Comput. Sci. Appl. **5** (2014)

Hawkins, R., Paterson, C., Picardi, C., Jia, Y., Calinescu, R., Habli, I.: Guidance on the Assurance of Machine Learning in Autonomous Systems (AMLAS). Assuring Autonomy International Programme (AAIP) University of York (Feb 2021)

Hein, D., Ivens, B., Müller, S.: Customer Acceptance Tests and New Product Success—An Application of QCA in Innovation Research, 2015/05/26 (2015)

Ihle, I.-A.F.: Software management for a critical, real-time product. In: Rolls-Royce Marine Dynamic Positioning Conference, 13–14 Oct 2015

Kent, J.: Test Automation from RecordPlayback to Frameworks. www.simplytesting.com/ (2007)

RINA: Rules for the certification of Quality Management System Certification Scheme for the Rail Sector ISO/TS 22163 Valid from 15/07/2020 (2020)

Ritchey, T.: General Morphological Analysis. A General Method for Non-quantified Modelling. Swedish Morphological Society (2013)

Stålhane, T., Myklebust, T.: Hazard stories, HazId and safety stories in SafeScrum. In: XP 2018, Porto, 2018

Tao, C., Gao, J., Wa, T.: Testing and quality validation for AI software perspectives, issues, and practices. IEEE Access. **7**, 120164–120175 (2019)

Tedder, D.: Four Things That Ensure "Fit for Use". https://itsm.tools/four-things-that-ensure-fit-for-use/ (2018)

Xie, X., Ho, J.W.K., Murphy, C., Kaiser, G., Xu, B., Chen, T.Y.: Testing and validating machine learning classifiers by metamorphic testing. J. Syst. Softw. **84**(4), 544–558 (2011)

Annex: Overview of Documents and Work Products Mentioned in Functional Safety Standards Including Weak Parts of Safety Standards

> *Tell me and I forget, teach me and I may remember, involve me and I learn.*
> —*Benjamin Franklin*

What This Chapter Is About
- Documents or work products mentioned in safety standards
- Documents not mentioned in safety standard series but often used by several manufacturers
- Weak parts of safety standards related to documentation and work products

A - Introduction

A typical proof of compliance (PoC) documentation consists of 50–200 documents (Myklebust et al., 2015). In addition, the safety standards have given titles to several documents: e.g., the IEC 61508:2010 series mentions 82 documents with a title, 101 documents in the EN 5012X series, and 106 work products (documents) in the ISO 26262:2018 series. The manufacturers normally reference the named documents in the safety case.

The referenced documents in the safety case also include references, typically 1–20 references. The total number of pages developed by the manufacturer is 2000–10,000 pages, depending on the project and the product/system to be developed.

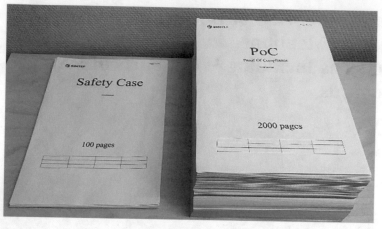

Fig. A.1 A safety case and the PoC (the referenced documents in the safety case). © Thor Myklebust

In this overview, we have limited the evaluated safety standards to the

- Generic safety standard IEC 61508 series
- Automotive safety standard ISO 26262 series
- Railway standard series EN 5012X

Relevant documents and weak or missing topics in current safety standards are described below.

B - Relevant Documents Not Mentioned in the Main Safety Standards

The safety standards have different approaches to documentation. While the railway standards use "documents," the automotive standard uses "work products." Other standards, such as the generic IEC 61508 series, are not specific but mention the topic in Annex A of IEC 61508-1:2010.

Documents not mentioned in these three standard series but often used by several manufacturers are:

1. Safety T&M to be used according to safety standards development projects. See, e.g., Annex in Part 2 and Part 3 of IEC 61508.
2. FAT, SAT, DAT, SIT, and CAT. FAT and SAT are mentioned in, e.g., IEC 61511, and described in IEC 62381:2012. In addition, IEC 62381:2012 includes requirements for SIT.
3. ODD/OEDR description (often part of the Operation envelope in the maritime industry).
4. SRAC document.

5. API (ISO 26262-8:2018 uses DIA (Development Interface Agreement), which have some similarities). Annex A of ISO 26262-8:2018 includes a DIA example.
6. FFFIS documents are mentioned in different railway TSIs.
7. Datasheet, see, e.g., edition 2018 of EN 50155 Annex J "Typical datasheet content."
8. Common cause report, see, e.g., NOROG 070 (2020).
9. Proof test procedures, see, e.g., NOROG 070 (2020).

C - Weak or Missing Parts in Safety Standards

The current editions of the safety standard have several weak or missing parts related to the presentation of relevant documents. Weak or missing topics in the three safety standards are listed and commented below. These topics will result in an extension in the list of relevant documents and information necessary.

Data Preparation

Data preparation is weak in all the three safety standards we have considered. However, data preparation has become more important in the last years due to the digitalization of safety systems – see chapter 8.4.7 EN 50128:2011 and Table A.11 "Data preparation Techniques." For further information and guidance, see SCSC guide (2021).

Safety Culture

It is important to have a safety-oriented work culture. Safety culture is defined in ISO 26262-1:2018 and included as a requirement in ISO 26262-2:2018 and ISO 26262-8:2018. However, safety culture is not mentioned in the current edition of the IEC 61508:2010 series.

Safety culture is not mentioned in the EN 5012X series. Safety culture is part of the Railway safety approach and is mentioned in the EU railway law [2016/798]. However, it mainly targets railway professionals (infrastructure management; carriage) as part of their internal safety organization more than manufacturers. EU and the EU agency for railways have issued a safety culture declaration (2021).

Processes/Life Cycles

Manufacturers should develop software according to the requirements of the relevant safety standards, but we propose an agnostic approach regarding life cycles. In the waterfall paper, Royce (1970) states, "*I believe in this concept, but the implementation described above is risky and invites failure*" Regarding safety standards and life cycle:

- IEC 61508-3:2018 states in clause 7.1.2.2: "*Any software lifecycle model may be used provided all the objectives and requirements of this clause are met.*" The next edition of this standard will be improved related to modern life cycles and will include more requirements and information related to incremental development and regression testing.

- EN 50128:2011 states in chapter 5.3.2.14, *"Where any alternative lifecycle or documentation structure is adopted, it shall be established that it meets all the objectives and requirements of this European Standard."* New life cycles for the safety domain, like SafeScrum, have also been developed (see Hanssen et al. (2018)).
- ISO 26262-6:2018 states: *"NOTE 1 Development approaches or methods from agile software development can also be suitable for the development of safety-related software, but if the safety activities are tailored in this manner, ISO 26262-2:2018 6.4.5 is considered. However, agile approaches and methods cannot be used to omit safety measures or ignore the fundamental documentation, process or safety integrity of product rigour required for the achievement of functional safety."*
- UL 4600:2020 has an agnostic approach to lifecycles.

Deployment

Deployment is mentioned neither in the IEC 61508:2010 series nor in the ISO 26262:2018 series. Deployment is mentioned in EN 50128:2011/A2:2020 but the standard does not include requirements or guidelines related to OTA (Over The Air) or DevOps. Deployment and maintenance are addressed in clause 9 of EN 50128.2011/A2:2020, and it appears that the authors of the standard have overlooked the fact that software may be developed in one project but deployed in another.

EN 50128:2011/A2:2020, clause 9.1.4.1 states: *"The deployment shall be carried out under the responsibility of the project manager."* Admittedly, the term *project manager* does not identify *which* project, but in real life, deployment of the final system by the customer might not necessarily be part of a project, in which case there will be no project manager in the sense of the definition given in clause 3 *"Terms, definitions and abbreviations."* The term *deployment* is not defined in clause 3 and is meant to correspond to what the other standards call installation and commissioning. Clause 9.1.4.13 states: *"A Deployment Record ... shall be stored among the delivered system related documents ... and is part of the commissioning and acceptance."* The requirements on software maintenance, described in clause 9.2, have been slightly modified. Notably, the requirement that *"Maintenance shall be performed with the same level of expertise, tools ... as the original development ...,"* stated in the previous edition's clause 16.4.5, has been dropped. Regular audits of maintenance activities against the software maintenance plan (old clause 16.4.4) and external supplier control (old clause 16.4.7) are also no longer addressed. On the other hand, verification of the plans and records is now required.

Distributed Development

Distributed development is a software development model in which software teams spread across geographical lines collaborate on developing a software system. The teams are often working separately and organized as mini-projects. Finally, the results are brought together for a final software release.

Distributed development is a familiar IT approach. In the last years, it has also become popular when developing safety-critical software. However, source code control and other issues make it less than ideal. Internet-based tools and collaboration software (e.g., Atlassian confluence) allow teams to work effectively as one distributed team.

The automotive safety standard series ISO 26262:2018 include definitions and requirements regarding distributed development. The definition is found in ISO 26262-1:2018. In addition, requirements and guidelines can also be found in ISO 26262-4:2011 "Product development," ISO 26262-6 "SW": 2018 and ISO 26262-8:2018 "Supporting processes."

ISO/IEC/IEEE 26515:2018, Systems and software engineering – "Developing information for users in an agile environment" includes a chapter "Globally distributed teams" that includes information related to the fact that agile principles may involve other considerations such as:

"Coordination of several information developers located across different development teams, and logistic issues and communication challenges related to different time zones and different languages."

This standard also includes chapter 6.3 regarding "Management of information development across teams using agile development".

Artificial Intelligence
Artificial Intelligence (AI) is the broad discipline of creating intelligent systems. Machine Learning (ML) refers to systems that can learn from experience. Deep Learning (DL) refers to systems that learn from experience on large data sets. Artificial Neural Networks (ANN) refers to models of human neural networks designed to help computers learn.

AI solutions, such as neural networks and machine learning, are fundamental to realizing the full potential of autonomous systems. However, the use of AI in safety-critical applications introduces challenges and well-established principles for safety verification break down. We need new knowledge on how to verify AI-enabled functionality and where the limitations are.

The functioning of a safety system involving AI is defined by pre-programmed logic and models that have been trained on massive amounts of data, resulting in a system that cannot be fully analyzed and comprehended in the traditional way, including safety verification. AI and, in particular, ML questions our perception of what software is, and we cannot anymore apply current functional safety standards alone.

D - Overview of Documents Mentioned in Safety Standards

IEC 61508:2010 series
The relevant documents are listed in IEC 61508-1:2010 Table D-A.1, D-A.2, and D-A.3. For some documents, it is possible to link the relevant requirement chapters

to the documents listed in A1–A3, and for other documents, this is not possible. Some documents have

- Links to a chapter
- Links to a T&M
- Links to a combination of the two above

The tables below have included the links between the documents and the relevant chapters and included comments when the links are missing in the standard series.

ISO 26262:2018 series

In this standard series, the work products are mentioned as deliverables concisely and constructively. The mentioned work products are listed in the table below (Sect. 3.2).

EN 5012X series

The three safety standards, EN 5012X, have a somewhat different approach and presentation of the relevant documents. All documents mentioned in EN 5012x are listed in the table below.

- EN 50126-1:2017: This safety standard has listed documents as deliverables in the different life cycle phases.
- EN 50128:2011: This standard summarizes the documents in Annex C "Documents control summary." They are divided into phases and responsibilities, first checker and second checker.
- EN 50129:2018: This standard is not as clear as the two previous standards regarding relevant documents or information. They are mentioned as logs specifications, etc.

Overview of IEC 61508:2010 Information/Documents

The amount of information documented may vary from a few lines to many pages, and the complete set of information may be divided and presented in several physical documents, compiled into one or more documents or, e.g., in a documentation system. The documentation structure will depend on the size and complexity of the E/E/PE safety-related systems and will take into account company procedures and the working practices of the specific product or application sector.

The example documentation structure indicated has been developed to illustrate how the information could be structured and how the documents could be titled. For "ready for production" documentation, see, e.g., ISO 26262-7:2018.

A document is a structured amount of information intended for human perception that may be interchanged as a unit between users and/or systems (see reference [16] in the Bibliography). Therefore, the document term applies not only to documents in the traditional sense but also to concepts such as data files and database information.

The term document is understood to mean information rather than physical documents unless this is explicitly declared or understood from the context of the clause or subclause in which it is stated. Documents may be available in different forms for human presentation (e.g., on paper, film, or any data medium to be presented on screens or displays).

The example documentation structure in this annex specifies documents in two parts:

- **Document kind**
- **Activity or object**

The document kind characterizes the content of the document, e.g., function description or circuit diagram. The activity or object describes the scope of the content, e.g., a pump control system. The basic document types specified in this annex are:

- **Specification** – specifies a required function, performance, or activity (e.g., requirements specification)
- **Description** – specifies a planned or actual function, design, performance, or activity (e.g., function description)
- **Instruction** – specifies in detail the instructions as to when and how to perform certain jobs (e.g., operator instruction)
- **Plan** – specifies the plan as to when, how, and by whom specific activities shall be performed (e.g., safety plan and maintenance plan)
- **Diagram** – specifies the function using a diagram (symbols and lines representing signals between the symbols)
- **List** – provides information in a list form (e.g., code list, signal list)
- **Log** – provides information on events in a chronological log form
- **Report** – describes the results of activities such as investigations, assessments, tests, verifications, validations, etc. (e.g., test report)
- **Request** – describes requested actions that have to be approved and further specified (e.g., maintenance request)

The basic document type may have a prefix, such as **requirements** specification or **test** specification, which further characterizes the content.

Table D-A.1 in IEC 61508-1 (as we expect ed.3 of IEC 61508-1 to be) – Example of a documentation structure for information related to the overall safety life cycle. © IEC

Overall safety life cycle phase	Information
Concept	Description (overall concept)
	Description (functional safety concept)
	Report (verification of the functional safety concept)
Overall scope definition	Description (overall scope definition)
Hazard and risk analysis	Description and report (hazard and risk analysis)
	Database or report (hazard log)

(continued)

Table D-A.1 (continued)

Overall safety life cycle phase	Information
Overall safety requirements	Specification (overall safety requirements, comprising: overall safety functions requirements and overall safety integrity requirements)
Overall safety requirements allocation	Description (overall safety requirements allocation)
Overall operation and maintenance planning	Plan (overall operation and maintenance)
Overall safety validation planning	Plan (overall safety validation)
Overall installation and commissioning planning	Plan (overall installation); Plan (overall commissioning)
E/E/PE system safety requirements	Specification (E/E/PE system safety requirements, comprising: E/E/PE system safety functions requirements and E/E/PE system safety integrity requirements)
E/E/PE safety-related system realisation	See Table D-A.2 and Table D-A.3
Overall installation and commissioning	Report (overall installation)Report (overall commissioning) Report (release) Report (deployment) Report (user manual)
Overall safety validation	Report (overall safety validation)
Overall operation and maintenance	Log (overall operation and maintenance)
Overall modification and retrofit	Request (overall modification);Report (overall modification and retrofit impact analysis);Log (overall modification and retrofit)
Decommissioning or disposal	Report (overall decommissioning or disposal impact analysis);Plan (overall decommissioning or disposal);Log (overall decommissioning or disposal)
Concerning all phases	Plan (safety);Plan (verification);Report (verification);Plan (functional safety assessment); Report (functional safety assessment)

Table D-A.2 in IEC 61508-1 (as we expect ed.3 of IEC 61508-1 to be) – Example of a documentation structure for information related to the E/E/PE system safety life cycle. © IEC

E/E/PE system safety life cycle phase	Information
E/E/PE system validation planning	Plan (E/E/PE system safety validation)
E/E/PE system design and development	
E/E/PE system architecture	Description (E/E/PE system architecture design, comprising: hardware architecture and

(continued)

Table D-A.2 (continued)

E/E/PE system safety life cycle phase	Information
	software architecture);Specification (programmable electronic integration tests);Specification (integration tests of programmable electronic and non-programmable electronic hardware); Instruction (development tools); Report (validation of tools);
Hardware architecture	Description (hardware architecture design); Specification (hardware architecture integration tests)
Hardware module design	Specification (hardware module design);Specifications (hardware module tests)
Component construction and/or procurement	Hardware modules;Report (hardware module tests)
Programmable electronic integration	Report (programmable electronic hardware and software integration tests) (see Table D-A.3)
E/E/PE system integration	Report (programmable electronic and other hardware integration tests)
E/E/PE system operation and maintenance procedures	Instruction (user);Instruction (operation and maintenance)
E/E/PE system safety validation	Report (E/E/PE system safety validation)
E/E/PE system modification	Instruction (E/E/PE system modification procedures);Request (E/E/PE system modification);Report (E/E/PE system modification impact analysis);Log (E/E/PE system modification)
Concerning all phases	Plan (E/E/PE system safety);Plan (E/E/PE system verification);Report (E/E/PE system verification);Plan (E/E/PE system functional safety assessment);Report (E/E/PE system functional safety assessment)
Concerning all relevant phases	Safety manual for compliant items

Table D-A.3 in IEC 61508-1 (as we expect ed.3 of IEC 61508-1 to be) – Example of a documentation structure for information related to the software safety life cycle. © IEC

Software safety life cycle phase	Information
Software safety requirements	Specification (software safety requirements, comprising: software safety functions requirements and software safety integrity requirements)
Software validation planning	Plan (software safety validation)

(continued)

Table D-A.3 (continued)

Software safety life cycle phase	Information
Software design and development Software architecture Software system design Software module design Coding Software module testing Software integration	Description (software architecture design) (see Table D-A.2 for hardware architecture design description); Specification (software architecture integration tests); Specification (programmable electronic hardware and software integration tests); Instruction (development tools and coding manual); Report (validation of tools); Description (software system design); Specification (software system integration tests); Specification (software module design); Specification (software module tests); List (source code); Report (software module tests); Report (code review); Report (software module tests); Report (software module integration tests); Report (software system integration tests); Report (software architecture integration tests)
Programmable electronic integration	Report (programmable electronic hardware and software integration tests)
Software operation and maintenance procedures	Instruction (user); Instruction (operation and maintenance)
Software safety validation	Report (software safety validation)
Software modification	Instruction (software modification procedures); Request (software modification); Report (software modification impact analysis); Report (software release); Report (software deployment); Log (software modification)
Concerning all phases	Plan (software safety); Plan (software verification); Report (software verification); Plan (software functional safety assessment); Report (software functional safety assessment)
Concerning all relevant phases	Safety manual for compliant items

Table D.1 Overview of ISO 26262:2018 work products

ISO 26262-2:2018 Management of functional safety	Overview ISO 26262-3:2018 Concept phase	Overview ISO 26262-4:2018 Product development at the system level	ISO 26262-5:2018 Product development at the hardware level
1. Organization-specific rules and processes for functional safety	1. Item definition	1. Technical safety requirements specification	1. Hardware safety requirements specification (including test and qualification criteria)
2. Evidence of competence management	2. Hazard analysis and risk assessment report	2. Technical safety concept	2. Hardware-software interface specification (HSI)
3. Evidence of quality management system	3. Verification report of the hazard analysis and risk assessment	3. System architectural design specification	3. Hardware safety requirements verification report
4. Identified safety anomaly reports	4. Functional safety concept	4. Hardware-software interface (HIS) specification	4. Hardware design specification (including test and evaluation criteria)
5. Impact analysis at the item level	5. Verification report of the functional safety concept	5. Specification of requirements for production, operation, service, and decommissioning	5. Hardware safety analysis report
6. Impact analysis at element level		6. Verification report for system architectural design, the hardware-software interface (HIS) specification, the specification of requirements for production, operation, service and decommissioning, and the technical safety	6. Hardware design verification report
7. Safety plan		7. Safety analysis report	7. Specification of requirements related to production, operation, service, and decommissioning
8. Safety case		8. Integration and test strategy	8. Analysis of the effectiveness of the architecture of the item to cope with the random hardware failures
9. Confirmation measure reports		9. Integration and test report	9. Verification review report of evaluation of the effectiveness of the architecture of the item to cope with the random hardware failures
10. Release for production report		10. Safety validation specification including safety validation environment description	10. Analysis of safety goal violations due to random hardware failures
11. Evidence of safety management regarding production, operation, service, and decommissioning		11. Safety validation report	11. Specification of dedicated measures for hardware
			12. Verification review report of evaluation of safety goal violations due to random hardware failures

(continued)

Table D.1 (continued)

ISO 26262-2:2018 Management of functional safety	Overview ISO 26262-3:2018 Concept phase	Overview ISO 26262-4:2018 Product development at the system level	ISO 26262-5:2018 Product development at the hardware level
			13. Hardware integration and verification specification 14. Hardware integration and verification report

ISO 26262-6:2018 Product development at the software level	ISO 26262-7:2018 Production, operation, service, and decommissioning	ISO 26262-8:2018 Supporting processes	ISO 26262-9:2018 Automotive Safety Integrity Level (ASIL)-oriented and safety-oriented analyses
1. Documentation of the software development environment 2. Software safety requirement specification 3. Hardware-software interface specification (refined) 4. Software verification report 5. Software architectural design specification 6. Safety analysis report 7. Dependent failure analysis report 8. Software verification report 9. Software unit design specification 10. Software unit implementation 11. Software verification specification 12. Software verification report 13. Embedded software 14. Configuration data specification	1. Safety-related content of the production plan 2. Safety-related content of the production control plan, including the test plan 3. Producibility requirements specification 4. Production process capability report 5. Safety-related content of the service plan 6. Safety-related content of the service instructions 7. Safety-related content of the information made available to the user 8. Safety-related content of the decommissioning instructions 9. Operation, service, and decommissioning requirement specification 10. Safety-related content of the rescue	1. Supplier selection report 2. Development interface agreement (DIA) 3. Suppliers safety plan 4. Functional safety assessment report 5. Supply agreement 6. Configuration management plan 7. Change management plan 8. Change request 9. "Impact analysis" and "change request plan" 10. Change report 11. Verification plan 12. Verification specification 13. Documentation management plan 14. Documentation guideline requirements 15. Software tool criteria evaluation report 16. Software qualification report 17. Software component documentation	1. Update of architectural information 2. Update of ASIL as attribute of safety requirements and elements 3. Update of ASIL as attribute of sub-elements of elements 4. Dependent failure analysis 5. Dependent failure analysis verification report 6. Safety analysis 7. Safety analysis verification report

(continued)

Table D.1 (continued)

ISO 26262-6:2018 Product development at the software level	ISO 26262-7:2018 Production, operation, service, and decommissioning	ISO 26262-8:2018 Supporting processes	ISO 26262-9:2018 Automotive Safety Integrity Level (ASIL)-oriented and safety-oriented analyses
15. Calibration data specification 16. Configuration data 17. Calibration data 18. Verification specification 19. Verification report	service instructions 11. Control measures report 12. Production process capability report 13. Field observation instructions	18. Software component qualification report 19. Software component qualification verification report 20. Hardware element evaluation plan 21. Hardware component test plan 22. Hardware element evaluation report for hardware elements 23. Description of candidate for proven-in-use argument 24. Proven-in-use analysis reports 25. Base vehicle manufacturer or supplier guideline 26. Safety rationale	

Table D.2 Overview of EN 5012X documents

EN 50126-1:2017	EN 50128:2011	EN 50129:2018
1. Independent safety assessment plan 2. Record of the independent safety assessment findings 3. Independent safety assessment report 4. Concept report (not given a name in the standard) 5. System definition 6. RAM plan 7. Safety plan 8. ConOps (if relevant, not concretized by the standard but mentioned operational conditions and context) 9. Risk assessment 10. Hazard log 11. RAMS requirement specification	1. Software quality assurance plan 2. Software quality assurance verification report 3. Software configuration management plan 4. Software verification plan 5. Software validation plan 6. Software requirements specification 7. Overall software test specification 8. Software requirements verification report 9. Software architecture specification 10. Software design specification 11. Software interface specifications	1. Definition of system 2. Safety plan 3. Hazard log 4. SRS 5. Description of system architecture and functional behavior (this can be a system architectural design specification refining the previous safety requirements specification) 6. Safety-related design principles 7. Description of interfaces 8. Modification procedures 9. Manufacturing documentation for this system 10. Applied documentation 11. Safety validation report 12. Software validation report 13. Software assessment report

(continued)

Table D.2 (continued)

EN 50126-1:2017	EN 50128:2011	EN 50129:2018
12. SRAC	12. Software integration test specification	14. Delivery sheet or release note
13. Validation for the phases 1–4	13. Software/hardware integration test specification	
14. Safety validation plan for the subsequent phases (5→)	14. Software architecture and design verification report	
15. Allocation of RAMS requirement specification to subsystems and/or components	15. Software component design specification	
16. Acceptance criteria and demonstration and acceptance processes and procedures	16. Software component test specification	
17. RAM analysis	17. Software component design verification report	
18. Hazard analysis	18. Software source code and supporting documentation	
19. Installation and commissioning procedures	19. Software source code verification report	
20. Operational and maintenance procedures	20. Software component test report	
21. Training measures	21. Software integration test report	
22. Plan(s) for future life cycle tasks	22. Software/hardware integration test report	
23. Safety case	23. Software integration verification report	
24. Quality assurance reports (regarding manufacturing process and measures for RAMS)	24. Overall software test report	
25. Inspection and testing reports	25. Software validation report	
26. Material handling and logistic arrangements	26. Tools validation report	
27. Installation documentation	27. Release notes (SW testing and final validation)	
28. Integration report	28. Application requirements specification	
29. Action taken to resolve failures and incompatibilities	29. Application preparation plan	
30. System support arrangements	30. Application test specification	
31. RAM validation report	31. Application architecture and design	
32. Safety validation report	32. Application preparation verification report	
33. ISA report	33. Application test report	
34. Endorsement of SRAC	34. Source code of application data/algorithms	
35. Acceptance report	35. Application data/algorithms verification report	
36. Plans and records suitable to trace the RAMS tasks undertaken within this life cycle phase (e.g., evidence of correct execution of the Operation and Maintenance Plan)	36. Software release and deployment plan	
37. Reports of RAMS	37. Software deployment	

(continued)

Table D.2 (continued)

EN 50126-1:2017	EN 50128:2011	EN 50129:2018
performance analysis and evaluations 　38. Identified recommendations and record of associated decisions 　39. Change request with impact analysis of the re-application of the system life cycle 　40. The impact on RAMS associated to the closing/dismantling of the system 　41. Decommissioning report	manual 38. Release notes (software) 39. Deployment records 40. Deployment verification report 41. Software maintenance plan 42. Software change records 43. Software maintenance records 44. Software maintenance verification report 45. Software assessment plan 46. Software assessment report	

References

- Directive 2016/798 OF THE EUROPEAN PARLIAMENT AND OF THE COUNCIL of 11 May 2016 on railway safety (recast)
- EU and European Union agency for railways. "The European Railway Safety Culture Declaration". The declaration can be downloaded at www.era.europa.eu/activities/safety-culture_en.Seen2021
- EN 50155:2017 Railway applications - Rolling stock - Electronic equipment
- G. K. Hanssen, T. Stålhane and T. Myklebust. SafeScrum – Agile Development of Safety-Critical Software. Springer December 2018
- IEC 62381:2012 Automation systems in the process industry – Factory acceptance test (FAT), site acceptance test (SAT), and site integration test (SIT)
- ISO/IEC/IEEE 26515:2018 Systems and software engineering – Developing information for users in an agile environment
- NOROG 070:2020 Remote and agile improvement of industrial control and safety systems processes N OIL AND GAS. Application of IEC 61508 and IEC 61511 in the Norwegian petroleum industry (Recommended SIL requirements)
- SCSC-127C: Data safety guidance Version 3.3. The data safety initiative working group (DSIWG), 2021

Printed in the United States
by Baker & Taylor Publisher Services